Process Development Division 2020

AA009247

Held at the 2020 AIChE Spring Meeting and 16th Global Congress on Process Safety

Online
17-21 August 2020

ISBN: 978-1-7138-2345-2

Printed from e-media with permission by:

Curran Associates, Inc.
57 Morehouse Lane
Red Hook, NY 12571

proceedings.com

AIChE
120 Wall Street, FL 23
New York, NY 10005-4020

Phone: (800) 242-4363
Fax: (203) 775-5177

www.aiche.org

Additional copies of this publication are available from:

Curran Associates, Inc.
57 Morehouse Lane
Red Hook, NY 12571 USA
Phone: 845-758-0400
Fax: 845-758-2633
Email: curran@proceedings.com
Web: www.proceedings.com

Process Development Division 2020

Held at the 2020 AIChE Spring Meeting and 16th Global Congress on Process Safety

Online
17-21 August 2020

AIChE
The Global Home of Chemical Engineers

TABLE OF CONTENTS

591161 2020 AIChE Spring Meeting & 16th GCPS AGILE Award Keynote Address - Beyond the Horizon: The Factors and Forces Affecting the Global Petrochemical Industry

Abstract:

Over the past 10 years, LyondellBasell and the global petrochemical industry has undergone a dramatic transformation. In this era of unprecedented disruption, it is likely that the next the decade will present both challenges and opportunities for the petrochemical sector. LyondellBasell CEO Bob Patel will summarize the company's progress to date and provide a forward-looking perspective on key opportunities and risks faced by the industry in 2020 and beyond.

591157 Securing Enterprises via Cybersecurity Understanding, Resiliency, and Education

Abstract:

Securing Enterprises via Cybersecurity Understanding, Resiliency, and Education

Industry 4.0 initiatives are enabled by leveraging the deployment of information technologies into the operational technology environment. While a logical approach in defining the future of smart manufacturing, it remains fret with challenges and risks that must be mitigated assure enterprise resilience to cyber and cyber-physical attacks. As economies recalibrate in the aftermath of the current public health emergency, definitions of what constitutes "critical infrastructure" are broadening while, at the same time, attack surfaces and the threat landscape are expanding. This means companies must place a greater emphasis on taking the steps and making the investments necessary to overcome the challenges and navigate these challenges and risks to maximize their resilience and secure and protect their enterprises. This presentation will explore the link between cybersecurity and process safety in the context of a safety culture. It will also identify cybersecurity challenges in industry 4.0 initiatives, the integration of information technologies (IT) and operational technologies (IT), the delivery of training, and the need for credentialing and ongoing professional development.

AN INTEGRATED PATTERN QUERY METHOD TO ASSIST PLANT MONITORING

Shu Xu, Anthony Amaro Jr., Mark Nixon
Emerson Automation Solutions

Introduction

Live plant behavior is usually recorded and saved as multivariate time series data which can be explored to assist process engineers in process monitoring, abnormality/fault detection and diagnosis. For example, patterns that indicate a trend or a fault might repeat itself over the whole operating period. In addition, faced with sufficient process data, process engineers are challenged to efficiently and effectively establish relationships between specific variables which share patterns with a high degree of correlation. In current work, we demonstrate the adoption of an integrated pattern query framework which combines PCA similarity factor and a fast similarity search algorithm based on dynamic time warping (DTW) to overcome such a challenge. Simulated industrial case studies illustrate the efficacy and efficiency of the proposed framework. Such results demonstrate the successful application of data analytics to cope with real problems in chemical manufacturing within the Digital Transformation era.

The remainder of this paper is organized as follows. Section 2 briefly introduces methodology information regarding PCA similarity factor, dynamic time warping, and an integrated framework. Section 3 provides case study details including the benchmark Tennessee Eastman process simulator and simulation conditions. Section 4 summarizes results obtained in the case study. We conclude the paper and discuss future works in Section 5.

Methods

PCA similarity factor

Principal component analysis (PCA) is a linear dimensional reduction technique that has been widely used in process monitoring applications [1,2]. PCA is intended to optimally capture data variations while minimizing the contaminating effect of random noise. In addition, PCA transforms the original correlated variables $\boldsymbol{x}_i$ into only a few independent principal components which are linear combinations of $\boldsymbol{x}_i$ and represent the directions of major variabilities. The PCA similarity factor metric [3,4] is proposed based on analyzing the structural similarity of such a transformed subspace. For a given two historical data sets, A and B, S_{PCA} can be calculated by:

$$S_{PCA}(A,B) = \frac{1}{k} trace(LM^T ML^T) = \frac{1}{k}\sum_{i=1}^{k}\sum_{j=1}^{k} cos^2\theta_{ij} \qquad (1)$$

where L and M are the matrices that contain the first k principal components of A and B, accounting for most of their variances respectively, and θ_{ij} is the angle between the ith principal component of A and the jth principal component of B. As we can see from Equation (1) the geometric interpretation of S_{PCA} is the sum of the squares of the cosines of the angles between each principal component of L and M.

Pattern-matching based on dynamic time warping

Dynamic time warping (DTW) is an advanced similarity measurement technique originally developed for speech recognition application, and it has been successfully applied to pre-processing chromatography data [5], mass spectroscopy (MS) data, and Nuclear Magnetic Resonance (NMR) spectrum data [6] and Batch Data[10]. In DTW, an optimization problem is formulated to maximize the geometrical similarity between two vectors, and it is commonly solved by dynamic programming. The mathematical framework for warping and alignment in DTW is introduced here. The nomenclature and the terminology follow the convention in Ref. [7].

First, we define reference trajectory R and target trajectory T of length N and M respectively as follows:

$$R := (r_1, r_2, \ldots, r_N), T := (t_1, t_2, \ldots, t_M) \quad (2)$$

We assume that the two trajectories are sampled at the same sampling interval. To measure the quality of alignment, a local cost function (or local distance measure) c is defined as:

$$c(r_n, t_m) = \|r_n - t_m\| \quad (3)$$

This definition assumes that r_n and t_m is of same dimension (which means same number of variables being measured).

Using the local cost function definition, we can define an alignment cost matrix $C \in \mathbb{R}^{N\times M}$

$$C(n, m) := c(r_n, t_m) \quad (4)$$

The goal of DTW is to find a warping path through the local cost matrix that gives us the minimal cumulative cost between the two trajectories. The warping path P is a series of pair-wise vectors that map sample indices from trajectory T to sample indices on R, and is defined as $P = (p_1, \ldots, p_L)$ and $p_l = (n_l, m_l) \in [1{:}N] \times [1{:}M]$. The warping trajectory is constrained by three conditions:

- Boundary condition: $p_1 = (1, 1)$, and $p_L = (N, M)$.
- Monotonicity condition: $n_1 \leq n_2 \leq \cdots \leq n_L$ and $m_1 \leq m_2 \leq \cdots \leq m_L$.
- Step-size condition: $p_{l+1} - p_l \in \{(1,0), (0,1), (1,1)\}$ for $l \in [1{:}L-1]$.

The boundary condition ensures that the first and the last samples map to the beginning and the end of the two trajectories. The monotonicity condition ensures that there will be no backtracking or "folding" of trajectories during warping. The step-size condition ensures that no sample indices will be skipped and limits the amount of warping applied to only three directions $(0, 1)$, $(1, 0)$ and $(1, 1)$.

Given any arbitrary warping path P, the total cost associated with the warping path and the two trajectories in question can be defined:

$$c_p(R, T) := \sum_{l=1}^{L} c(r_{nl}, t_{ml}) \quad (5)$$

With these definitions in place, we can formulate the objective function for DTW warping. The optimal warping path determined through DTW aims to minimize the total cost $c_p(R, T)$.

$$p^*_{DTW} := \arg\min\{c_p(R, T)\} \quad (6)$$

It could be very difficult to find p^* for long trajectories since the search space would be of dimension $\mathbb{R}^{N\times M}$. An efficient procedure to solve this problem in polynomial time $O(M \cdot N)$ based on dynamic programming has been devised.

From the local cost matrix $C(R, T)$, a cumulative cost matrix can be defined as:

$$D(n, m) := DTW(R(1{:}n), T(1{:}m)) \tag{7}$$

Where each element in the matrix D represents the cumulative cost accrued arriving at (m, n) from the starting point $(1, 1)$ by taking the optimal path subjected to the three warping constraints.

The elements of the matrix D can be filled out efficiently using the following relationship:

$$D(n, 1) := \sum_{k=1}^{n} c(r_k, t_1), n \in [1{:}N] \tag{8}$$

$$D(1, m) := \sum_{k=1}^{m} c(r_1, t_k), m \in [1{:}M] \tag{9}$$

$$D(n, m) := \min\{D(n-1, m-1), D(n-1, m), D(n, m-1)\} + c(r_n, t_m) \tag{10}$$

To perform DTW, the cumulative cost matrix D can first be calculated using the rules shown in Equations (8) ~ (10). It takes $O(M \cdot N)$ iterations to populate the entire cumulative cost matrix. After the D matrix is populated, the optimal warping trajectory p^*_{DTW} can be found by back-stepping from the end (N, M) back to the starting point $(1, 1)$ using the following rule.

$$p_{l-1} := \begin{cases} (1, m-1), & n = 1 \\ (n-1, 1), & m = 1 \\ \arg\min\{D(n-1, m-1), D(n-1, m), D(n, m-1)\}, & otherwise \end{cases} \tag{11}$$

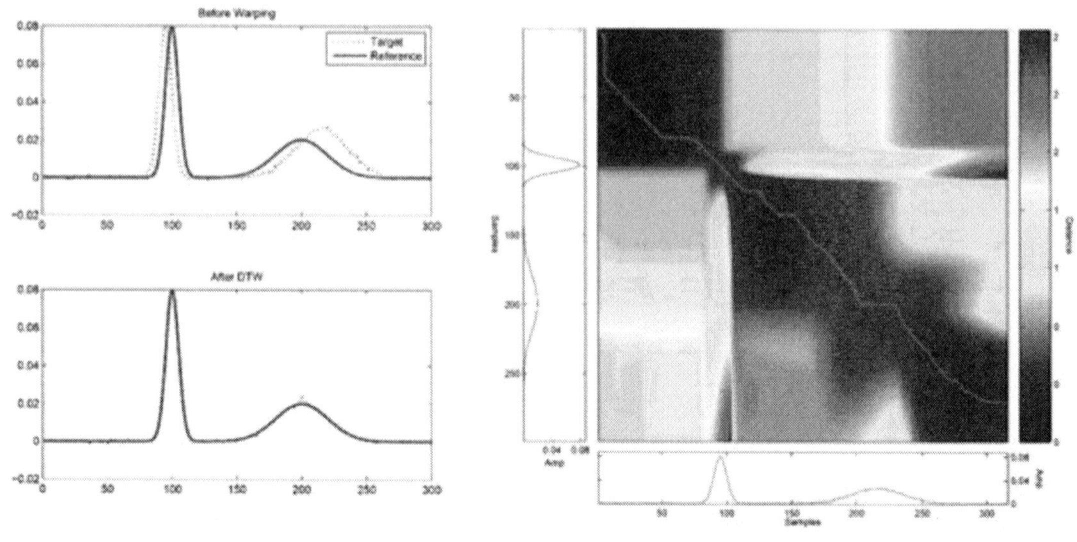

(a) Trajectories from before (top) and after (bottom) warping

(b) The cumulative distance matrix D (heat map) and the optimal path (purple line)

Figure 1 Trajectory plot and diagnostics from before and DTW

Figure 1 shows a result of DTW algorithm on a simple two-trajectory warping problem, where Figure 1(a) shows that the target trajectory (dashed line) has been warped to conform the reference (red line), and Figure 1(b) illustrates the cumulative cost matrix D for this problem. The darker and more blue color indicates areas of lower cumulative cost. The purple line indicates the optimal path that achieves the lowest cumulative cost as defined in Equation (11).

It is interesting to note that the (n,m)th element of the cumulative cost matrix D, $D(n, m)$, can be used as an index to measure the similarity between two univariate signals, and can assist process engineers in pattern matching and identification.

An integrated methodology framework

In this paper, we propose an integrated methodology which combines the PCA similar factor with the DTW-based pattern matching. The flow chart of such a method is illustrated in Figure 2. Given a multivariate data set ***A*** containing a pattern indicating a trend or a fault, we first extract the structure of the data set ***A***, and use it to find periods of data with a high PCA similar factor. Then we identify key contributors to such a trend or a fault via variable contribution plots. Last, we refine our searching based on the DTW-based pattern matching technique, as discussed in the section shown above.

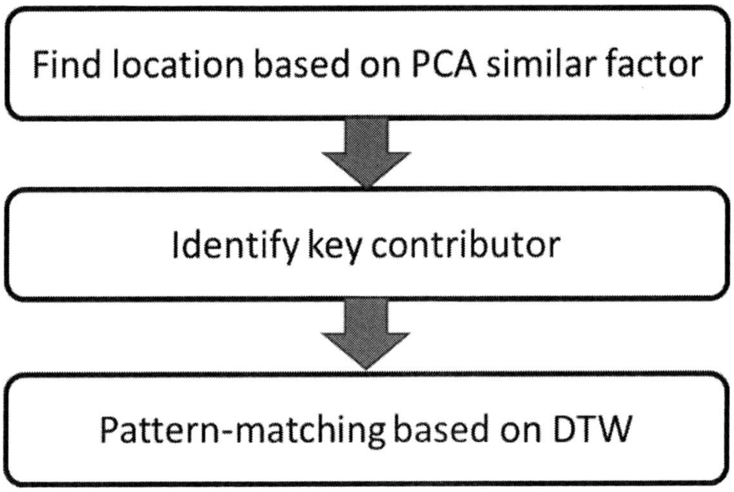

Figure 2 Flow chart of an integrated methodology of pattern query

Case study

The benchmark Tennessee Eastman process simulator

The Tennessee Eastman process model is a realistic simulation program of a chemical plant which is widely accepted as a benchmark for evaluating process control and monitoring methods [1]. Figure 3 shows the flow diagram of the process with five major units, i.e., a reactor, condenser, compressor, separator and stripper. The process has two products from four reactants. Additionally, an inert and a by-product are also present making a total of 8 components denoted as A, B, C, D, E, F, G and H. The process allows for a total 52 measurements out of which 41 are of process variables and 11 are manipulated variables, see Tables 1 and 2:

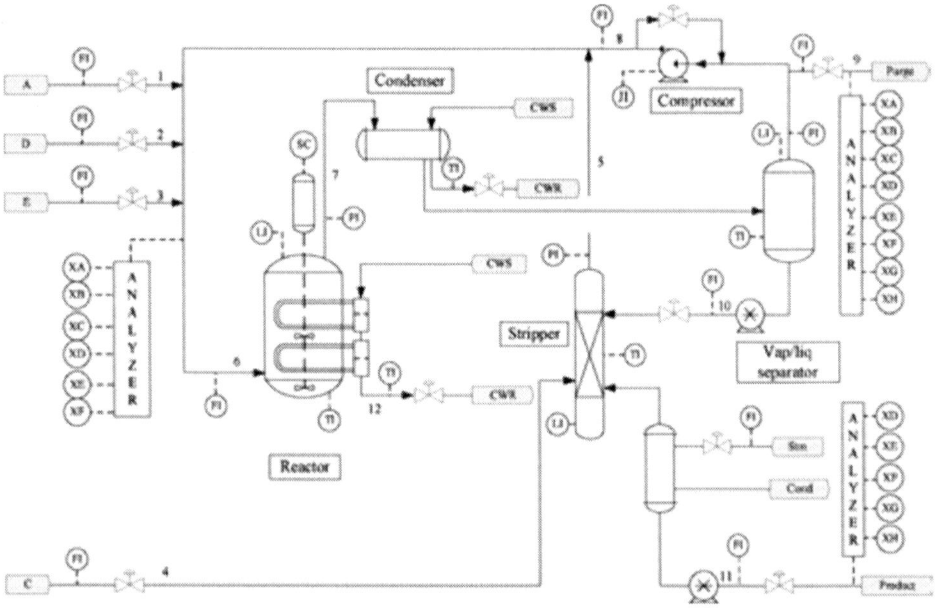

Figure 3 The Tennessee Eastman process

Table 1 Process variables

Block name	Variable name	Number
Input feed	A feed (stream 1)	XMEAS(1)
	D feed (stream 2)	XMEAS(2)
	E feed (stream 3)	XMEAS(3)
	A and C feed	XMEAS(4)
Reactor	Reactor feed rate	XMEAS(6)
	Reactor pressure	XMEAS(7)
	Reactor level	XMEAS(8)
	Reactor temperature	XMEAS(9)
Separator	Separator temperature	XMEAS(11)
	Separator level	XMEAS(12)
	Separator pressure	XMEAS(13)
	Separator underflow	XMEAS(14)
Stripper	Stripper level	XMEAS(15)
	Stripper pressure	XMEAS(16)
	Stripper underflow	XMEAS(17)
	Stripper temperature	XMEAS(18)
	Stripper steam flow	XMEAS(19)
Miscellaneous	Recycle flow	XMEAS(5)
	Purge rate	XMEAS(10)
	Compressor work	XMEAS(20)
	Reactor water temperature	XMEAS(21)
	Separator water temperature	XMEAS(22)

Block name	**Variable name**	**Number**
Reactor feed analysis	Component A	XMEAS(23)
	Component B	XMEAS(24)
	Component C	XMEAS(25)
	Component D	XMEAS(26)
	Component E	XMEAS(27)
	Component F	XMEAS(28)
Purge gas analysis	Component A	XMEAS(29)
	Component B	XMEAS(30)
	Component C	XMEAS(31)
	Component D	XMEAS(32)
	Component E	XMEAS(33)
	Component F	XMEAS(34)
	Component G	XMEAS(35)
	Component H	XMEAS(36)
Product analysis	Component D	XMEAS(37)
	Component E	XMEAS(38)
	Component F	XMEAS(39)
	Component G	XMEAS(40)
	Component H	XMEAS(41)

Table 2 Process manipulated variables

Variable name	Number	Base value	Units
D feed flow	XMV(1)	63.053	$kg\,h^{-1}$
E feed flow	XMV(2)	53.980	$kg\,h^{-1}$
A feed flow	XMV(3)	24.644	kscm h
A and C feed flow	XMV(4)	61.302	kscm h
Compressor recycle valve	XMV(5)	22.210	%
Purge valve	XMV(6)	40.064	%
Separator pot liquid flow	XMV(7)	38.100	$m^3\,h^{-1}$
Stripper liquid product flow	XMV(8)	46.534	$m^3\,h^{-1}$
Stripper steam valve	XMV(9)	47.446	%
Reactor cooling water flow	XMV(10)	41.106	$m^3\,h^{-1}$
Condenser cooling water flow	XMV(11)	18.114	$m^3\,h^{-1}$
Agitator speed	XMV(12)	50.000	rpm

Downs and Fogel [8] initially defined 20 process faults and an additional valve fault was further introduced in [1], see Table 3.

Table 3 Descriptions of process faults in TE process

Fault number	Process variable	Type
IDV(1)	A/C feed ratio, B composition constant	Step
IDV(2)	B composition, A/C ration constant	Step
IDV(3)	D feed temperature	Step
IDV(4)	Reactor cooling water inlet temperature	Step
IDV(5)	Condenser cooling water inlet temperature	Step
IDV(6)	A feed loss	Step
IDV(7)	C header pressure loss-reduced availability	Step
IDV(8)	A, B, and C feed composition	Random variation
IDV(9)	D feed temperature	Random variation
IDV(10)	C feed temperature	Random variation
IDV(11)	Reactor cooling water inlet temperature	Random variation
IDV(12)	Condenser cooling water inlet temperature	Random variation
IDV(13)	Reaction kinetics	Slow drift
IDV(14)	Reactor cooling water valve	Sticking
IDV(15)	Condenser cooling water valve	Sticking
IDV(16)	Unknown	Unknown
IDV(17)	Unknown	Unknown
IDV(18)	Unknown	Unknown
IDV(19)	Unknown	Unknown
IDV(20)	Unknown	Unknown
IDV(21)	The valve fixed at steady state position	Constant position

Simulation conditions

The data set generated using the TE simulator above have been widely used in process monitoring studies; for example Rieth et al. generated both training and testing data for fault detection evaluation [9]. By considering the time constants of the process in closed loop, the sampling time was selected as 3 minutes. In our study, process data are generated containing 21 randomly generated process faults described in Table 3. A moving window approach is employed to calculate the PCA similarity index and the DTW-based distance, where the window contains a pattern indicating a trend or a fault.

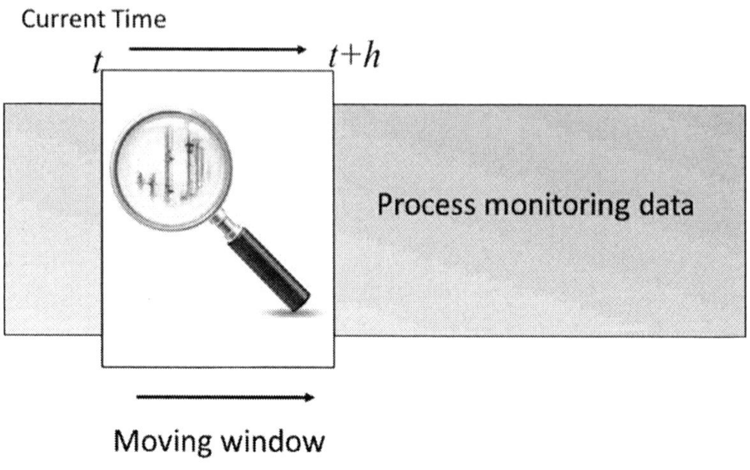

Figure 4 A moving window approach for pattern query

Results and Discussions

Among different types of faults, two of them have been selected for demonstration: a step change (Fault #1) and a sticking valve (Fault #14). The pattern matching results for Faults #1 and #14 are illustrated in Figures 5 and 6. As shown in Figures 5(a) and 6(a), the peaks associated with the locations of Fault #1 and #14 are highlighted in red squares. It is observed that Fault #1 occurs more regularly than Fault #14, and its PCA similarity indices enjoy a smaller standard deviation. After locating where the faults are, we identify the top contributors to both faults through variable contribution plots, and pinpoint their individual patterns using the dynamic time warping (DTW)-based pattern-matching approach, as shown in Figures 5(b) and 6(b). It is observed that the DTW aligns the peaks and valleys of the signal well despite the presence of signal noise. In addition, the Euclidean distance metric used in evaluating the pattern similarity between the two signals obtains a satisfactory result.

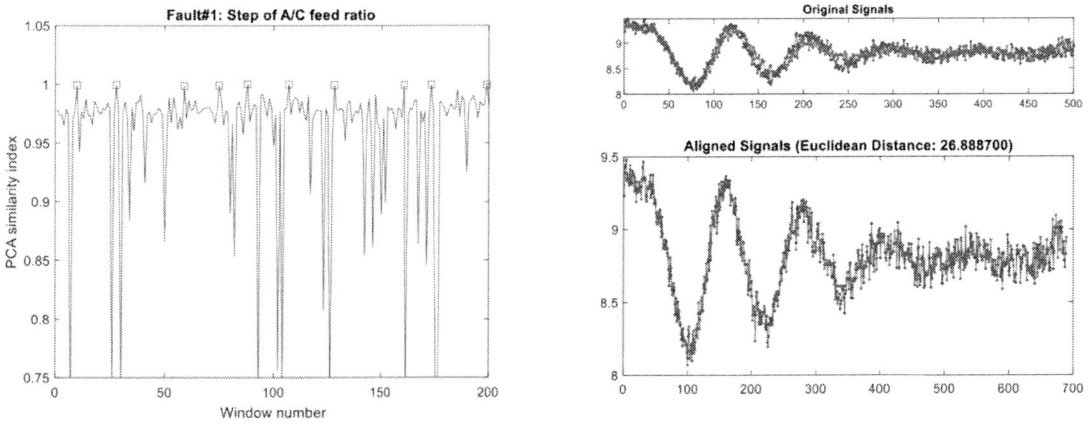

(a) Location found for Fault#1 based on PCA similar index

(b) Pattern-matching based on DTW

Figure 5 Pattern query results for Fault #1

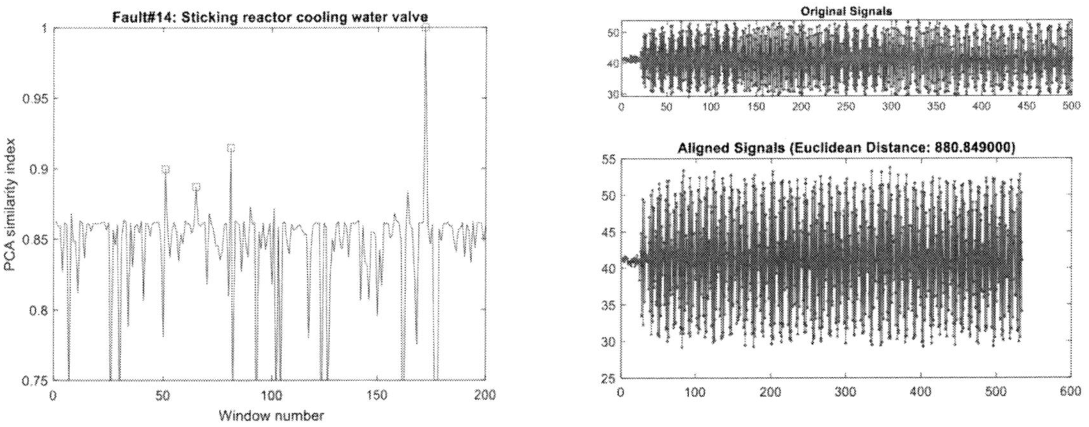

(a) Location found for Fault#14 based on PCA similar index

(b) Pattern-matching based on DTW

Figure 6 Pattern query results for Fault #14

Conclusions and Future Work

In the current work, we demonstrate the adoption of an integrated pattern query framework to overcome the challenges faced in process monitoring and fault detection. Such a framework combines PCA similar factor and dynamic time warping (DTW)-based pattern matching approaches, and is tested on data generated from the benchmark Tennessee Eastman process simulator. The method successfully finds periods where process faults occur and pinpoints the measurements of key contributing variables which are similar to the established pattern. Such results illustrate the successful application of data analytics to wrestle with real problems in chemical manufacturing within the Digital Transformation era.

In the future, we will test such a pattern query framework on more industrial data and compare it with other methods.

References

[1] Chiang, L.H., Russell, E.L., and Braatz, R.D. (2000). *Fault Detection and diagnosis in industrial systems*. Springer Science & Business Media.

[2] Kourti, T. and MacGregor, J.F. (1996). "Multivariate SPC Methods for Process and Product Monitoring," *J. Quality Tech.*, 28, 409-428.

[3] Krzanowski, W.J. (1979). Between-Groups Comparison of Principal Components. *J. Amer. Stat. Assoc.*, 74(367), 703-707.

[4] Singhal, A. and Seborg, D.E. (2001) "Matching patterns from historical data using PCA and distance similarity factors," *In Proceedings of the 2001 American Control Conference*. Vol. 2, 1759-1764. IEEE.

[5] Tomasi, G., Van Den Berg, F., Andersson, C.(2004). "Correlation optimized warping and dynamic time warping as preprocessing methods for chromatographic data," *J. Chemom.*, 18, 231-241.

[6] Torgrip,R.J.O, Åberg,M., Karlberg,B., Jacobsson, S.P.(2003). "Peak alignment using reduced set mapping," *J. Chemom.*, 17, 573-582, 2003.

[7] Müller, M. (2007). *Information retrieval for music and motion*.

[8] Downs, J. and Fogel, E.(1993). "A plant-wide industrial process control problem," *Comput. Chem. Eng.* 17, 245-255.

[9] Rieth, C.A.,Amsel, B.D., Tran, R. and Cook, M.B.(2017). "Additional Tennessee Eastman Process Simulation Data for Anomaly Detection Evaluation," *Harvard Dataverse*.

[10] Zhang, Y., Lu, B., and Edgar, T. F. (2013). "Batch Trajectory Synchronization with Robust Derivative Dynamic Time Warping", *ACS Publications*.

589663 Machine Learning Assisted Preventive Maintenance

Abstract:

Machine Learning Assisted Preventive Maintenance

Brian Koren

Graduate Student

Swalm School of Chemical Engineering,

Mississippi State University

bjk187@msstate.edu

Amin Amirlatifi

Assistant professor

Swalm School of Chemical Engineering,

Mississippi State University

amin@che.msstate.edu

ABSTRACT

Majority of manufacturing and chemical plants rely on the knowledge of experienced maintenance staff to plan, schedule, implement, and even predict routine and emergency maintenance needs to avoid unplanned downtime. As many of these knowledgeable staff members begin to reach retirement age, there is a growing demand to document, standardize, and accurately time preventive maintenance, instead of following a predetermined time schedule. This demand is increased by the growing availability of different techniques, such as ultrasonic, vibration and heat analysis, to help predict failures.

The staggering amount of data for predicting maintenance has prompted this research by creating a need for a suite of machine learning methods to analyze and utilize all the different inputs available to predict failures. The present study uses natural language processing (NLP) to read scheduled maintenance plans and analyze maintenance staff notes on how equipment has failed, and what corrective measures were taken afterwards. This information is coupled with time series analysis of data from the time leading up to that failure from the ensemble of sensors in the plant, to define key predictors of the failure using Self Organizing Maps (SOMs) and k-Nearest Neighbors (KNN). A deep neural network (DNN) is trained on the observed trends of key predictors using a combination of Long Short Term Memory (LSTM) and Convolutional Neural Networks (CNN) and system health (healthy/fault prone/failing/failed). The trained system is then used against real-time data from the sensors to predict an upcoming failure, its severity, and prescribe possible remedial actions ahead of failure.

The results from this project showed that many common equipment failures can be defined and predicted by coupling together data from many different inputs surrounding the equipment. The prediction accuracy also improves as more data becomes available to train the model, and when more sensors and inputs are available. The results of this project should encourage industry reliability teams to increase the sensors available for critical equipment and to apply machine learning techniques to handle all the inputs for improved failure prediction.

590246 Data Analytics - a Roadmap to Achieve Operational Excellence

Abstract:

Asset Optimization and Operational performance can be improved significantly by using advanced analytical tools such as Big Data Analytics, IIoT/P, and AI in combination with fundamental models by developing a "**Digital Twin**".

Ingenero uses the "**Digital Twin**" approach to identify operational gaps and effective resolution of issues in timely manner to maximize performance. Ingenero's *Excellence Through Insight* program is successful through combining the right information with in-depth domain expertise and deep data analytics to generate predictive and prescriptive recommendations presented in customized visual displays and dashboards for intelligent decision making.

The success of the process requires the appropriate volume of data. Post collection and mining of data, tool are used for effective reconciliation, validation and consolidation of data to gain insights quickly. This information is utilized as an input for machine learning models and these models convert the raw information into useful statistics/analytics that maximize asset effectiveness.

This paper will discuss Digitalization process utilizing fundamental models combined with data analytic techniques to achieve operational excellence. The appropriate techniques, tools and equipment models that are applied to track the performance and identify the size of deviation and improvement opportunities on real-time basis will be discussed. Additionally the paper will describe how predictive and prescriptive analytics can help determine the impact of deviation, need for corrective action and assist in reaching challenging solutions.

The Industry 4.0 Digitalization initiative is driving industry specific advanced analytics to leverage both historic and real-time data to achieve improvements that are difficult to accomplish using traditional approaches. Substantial economic gains can be achieved when data analytics is coupled with the overall plant operating strategy.

Application of right data analytics process has helped manufacturers to improve their Overall Plant Performance and Capacity by 5 – 15%.

590181 A Critical Review with Compilation of Pure and Multicomponent Gas Adsorption Equilibria and Isosteric Heat of Adsorption Data

Abstract:

We present a compilation and critical review of the existing literature data on gas adsorption equilibria and isosteric heat of adsorption. The NIST/ARPA-E adsorption database[1] covers wide ranges of pure component adsorption isotherms. However, it does not cover data for mixed-gas adsorption equilibria nor isosteric heat of adsorption. These data are essential for the development of robust multicomponent gas adsorption equilibria model such as adsorption non-random two liquid (aNRTL). The reliable predictions of the models depends on the quality of experimentally measured data and experimental measurement techniques. In this work, we compile and analyze the quality of mixed-gas adsorption experimental data and uncertainty associated with it. We further present prediction of the selected models for several multiple component systems.

References

1. Siderius, D. W.; Shen, V. K.; Johnson III, R. D.; van Zee, R. D., *NIST/ARPA-E Database of Novel and Emerging Adsorbent Materials, National Institute of Standards and Technology.* Gaithersburg, MD, 20899, https://dx.doi.org/10.18434/T43882, (retrieved August 07, 2019).

589847 Simulation of Pressure Swing Adsorbers Using the Moving Bed Model

Abstract:

Simulation of pressure swing adsorption (PSA) units remains a challenge for practicing engineers, which hinders the process intensification potential of the technology. The Moving Bed model, a steady-state model which represents any adsorption process as a series of interconnected continuous adsorption units, has the potential to alleviate some of the difficulties in modeling PSA. [1, 2, 3] In the Moving Bed model, the cycle is divided into multiple constant-pressure operating states bearing similarity to the individual steps of the cycle which allows the properties of the state to be evaluated from equilibrium considerations alone. The operating steps are arranged from highest-to-lowest pressure to represent the driving force gradient exploited during the cycle and then connected by a virtual moving bed. The virtual moving bed does not physically exist in the system but represents the adsorbate and void space hold-up retained in the adsorption column at the end of the operating state. However, modeling a cyclic batch process as a continuous, equilibrium process neglects the rate-based transport phenomena involved. Using the idea of adsorption efficiencies [4] to account for non-idealities like mass transfer resistance and dispersion, similar to the tray efficiencies used in rate-based distillation modeling, the retentate flow rates can be adjusted to match the actual performance of the unit. [2, 3] To showcase the capabilities and current limitations of the model, a case study will be presented which investigates a lab-scale air separation unit. Guidelines for ensuring convergence and robustness will be given and the effect of important parameters will be discussed.

1. Sees, M.D. (2018, October). *Novel Steady-State Process Modeling Methodology for Pressure Swing Adsorption.* Paper presented at the 2018 Annual Meeting of the American Institute of Chemical Engineers, Pittsburgh, PA.
2. Kirkes, T. (2018, October). *Steady-State Modeling of Air Separation by Pressure Swing Adsorption Using a Bed-State Efficiency Model.* Poster presented at the 2018 Annual Meeting of the American Institute of Chemical Engineers, Pittsburgh, PA.
3. Sees, M.D. (2019, April). *Modeling of Gas Separations by Pressure Swing Adsorption Using a Novel Steady-State Methodology.* Paper presented at the 2019 Spring Meeting of the American Institute of Chemical Engineers, New Orleans, LA.
4. Smith, O.J. and A.W. Westerberg. *Eng. Sci.*, 1991, 48(12), 2967 – 2976.

589746 Simulation of Near-Optimal and Valve-Free Operational-Half of Pressure Swing Adsorption Processes for Fair and Facile Evaluation of Process Performance Metrics

Abstract:

In this contribution, we will present a method for simulating *near-optimal* and *valve-free operational-half* of pressure swing adsorption (PSA) processes based on our new modeling and simulation framework, so called the *valve-free model*. The framework automatically determines *near-optimal* operational decisions during dynamic simulation of a PSA process. Also, the model eliminates any necessity for the user to specify inputs related to the detailed operation of a PSA process. Therefore, by using the framework, any non-expert users can easily obtain simulation results corresponding to a *near-optimal* operation of a PSA cycle.

First principle dynamic modeling and simulation of PSA processes has widely been used to capture dynamics of non-equilibrium adsorption phenomena. Furthermore, the approach has potentials to be utilized in many industrially relevant applications such as adsorbent screening, conceptual process design, optimal control, and optimization. Despite these potentials, however, there have been several limitations that prevented widespread adoptions of the first principle modeling and simulation in providing solutions to the aforementioned applications.

One of the major limitations of the current standard dynamic model, i.e. *valve* model, is its mathematically intractable formulation. For instance, in order to mathematically describe an adsorption column, the model needs to be formulated using a coupled system of nonlinear partial differential equations. Typically, solution methods to a such model are quite involved in terms of numerical methods used, conceptually hard to understand, and difficult to implement.

Another major limitation of the *valve* model is related to the user experiences in utilizing the model. Currently, to simulate a *near-optimal* PSA process using the *valve* model, the user has to supply a carefully chosen set of input parameters for a given PSA process simulation. However, choosing a set of input parameters that leads to *near-optimal* process simulation is not so simple as there practically exists so many reasonable ways to specify inputs such that the final simulation results are nowhere *near-optimal*.

Lastly, one of the most critical limitations of the *valve* model is the apparent sensitivity of simulation outputs on user supplied inputs. When designing chemical process systems, having accurate process performance metrics is considered as an essential requirement in determining process feasibility. However, due to the sensitivity issue, performance metrics obtained by the *valve* model are substantially influenced by the user supplied inputs. Furthermore, identifying an optimal set of inputs for *valve* model is computationally exhausting as it requires time consuming *trial-and-error* simulation studies.

Therefore, these limitations prevented a broader group of researchers and practitioners from being able to accurately predict performance metrics related to PSA process. Since enabling chemical industry to reliably assess adsorption based *process intensification* (PI) concepts would lead to a faster and more widespread deployment of such novel technologies, there is a critical need for developing a new dynamic model for simulating *near-optimal* PSA processes.

To address this critical need, a novel *valve-free* model has been developed to enable the user to simulate *near-optimal* PSA processes with *Skarstrom* type PSA cycles. In order to develop the *valve-free* model, the *Skarstrom* PSA cycle was decoupled into two halves including different steps, namely the *operational-half* and *re-generation-half*. Based on the simplified system, *near-optimal* operational policies for the two halves were devised separately.

A *near-optimal* operational policy for the *re-generation-half* was presented in previous talks given during the 2019 Spring AIChE meeting[1] as well as the 2019 Annual AIChE meeting[2]. Therefore, in this oral presentation, we are going to discuss the technical details that are involved in the development of a *near-optimal* operational policy for the *operational-half*.

In order to develop the *valve-free* model, several key assumptions were made to simplify the system and to present a *proof-of-concept* for the proposed modeling framework. The current formulation of the *valve-free* model assumes isothermal, no dispersion effects, binary ideal gas mixture, linear isotherm, and linear driving force adsorption kinetics.

The power of the proposed methodology (i.e. *valve-free* model) lies within enabling the user to run high quality PSA process simulations without having to preconceive the respective optimal operational policies. This enhanced user experience may lead to more wide adoption of process simulation of PSA processes as a tool that can help make critical day-to-day decisions or assessments that are related to operation, control, and optimization of PSA processes.

For the scope of the future research goals, the merge between the decoupled halves will be sought after in hoping to devise an overall *near-optimal* operational policy for PSA processes involving *Skarstrom* type cycles. Sequentially, based on the *near-optimal* operational policy for PSA processes generated by our *valve-free* model, further interrogations on the development of design rules for synthesizing *near-optimal* PSA processes will be made.

Ultimately, by serving as practical tools that can address the identified knowledge gap, *valve-free* model can potentially contribute to the advancement of PI and deployment of various adsorption technology in the chemical process industry.

[1] Kim, T.-H. et al. (2019). Rational Screening of Adsorbents for Natural Gas Upgrading By Pressure Swing Adsorption Using Dynamic Simulation of Process Performance Metrics, April 2, 2019, Spring AIChE Meeting, New Orleans, LA, 2019.

[2] Kim, T.-H. et al. (2019). A 'Valve-Free' Model for Dynamic Simulation of Pressure Swing Adsorption Processes That Automatically Determines Near-Optimal Operational Policies, November 11, 2019, Annual AIChE Meeting, Orlando, FL, 2019.

591338 Supercritical Fluid Extraction for Process Intensification of Biological Processes

Abstract:

The energy efficiency of many biological processes is limited by recovery of dilute fermentation products. Ethanol is an exception to this rule, as it can be tolerated by microbes at concentrations greater than 10 wt%. Other products are much more toxic, which limits their endpoint concentrations and makes energy intensive their recovery. While distillation remains the separation technology of choice when feasible, energy efficient alternatives are required for instances where endpoint concentrations are insufficient for distillation, when products are insufficiently volatile, or when products are thermally sensitive. One promising technology, which has been known since the 1980s but incompletely developed, is supercritical fluid extraction. Supercritical fluid extraction combines elements of distillation with liquid-liquid extraction. The extraction solvent, typically carbon dioxide or a light hydrocarbon, is present in its supercritical state, meaning that the thermodynamic driving force is based on solute partitioning – similar to liquid-liquid extraction. Subsequent to extraction, the target molecule can be recovered in high purity simply by decreasing the pressure, meaning that supercritical fluid extraction does not require downstream separation steps – a feature that is held in common with distillation. Ideal targets should be amenable to biomanufacturing, but with titers limited by end product inhibition. They should be water soluble at concentrations less than their inhibition limit and partition favorably into one or more supercritical solvents and out of aqueous solution. Pressure and temperature can be adjusted to fine tune the separation. Butanol is used as a test case to show dramatic, >10-fold decreases in energy requirements for separation. New advances that combine microbes that are tolerant of supercritical fluids and capable of bioproduction of desirable target molecules hold promise for even greater improvements in energy efficiency, as demonstrated by a recent study using the newly identified strain *Bacillus megaterium SR7*. Integrated fermentation and extraction benefits from optimized process design including solvent recycling and co-production of carbon dioxide at high pressure. Future studies can investigate mass transfer in the supercritical fluid-water environment. Future development work can focus on reducing costs.

591331 Optimizing Cyclic PSA/TSA Conditions for Ammonia Separation with Supported Metal Halides

Abstract:

The recent push towards reduction of carbon dioxide emissions has been the main reason for finding alternatives to the conventional Haber-Bosch ammonia production process. Although well-optimized through a century of research and development, the high operating pressure of the Haber-Bosch process is economically demanding. Alternative pathways to produce green ammonia introduce renewable energy (mostly wind-generated) as a base for ammonia production with the downside being double the cost. This shortcoming is mitigated by operating the process at lower pressure by replacing the condenser with an absorber containing metal halides capable of better ammonia separation – using the reaction-absorption process.[1–4] The application of absorption for ammonia separation is well studied; yet, there is no information about the rates of ammonia release from the metal halide absorbents.[5,6]

In this presentation, we offer one of the initial efforts to study ammonia release from different metal halide absorbents. The target is to better understand the ammonia release rates and absorbent working capacity under the cyclic pressure and temperature swing absorption conditions (PSA/TSA). A lab-scale absorber packed with supported and unsupported materials is used to observe ammonia breakthrough and release times for each cycle. Ammonia release at specific temperatures and pressures is recorded in carefully controlled experiments. Absorption tests were conducted at 40 psi with temperatures in the range of 25-400 °C. Results show a viable path to achieve reproducible breakthrough curves while achieving ammonia concentrations above 90%. The effect of the sweep gas during the desorption is also studied with the purpose of finding the optimal values for sweep time and volumetric flow. These findings will guide us to further optimize the small reaction-absorption ammonia production process and explore the strategies for further scale-up.

References:

(1) Malmali, M.; Wei, Y.; McCormick, A.; Cussler, E. L. Ammonia Synthesis at Reduced Pressure via Reactive Separation. *Ind. Eng. Chem. Res.* **2016**, *55* (33), 8922–8932.

(2) Malmali, M.; Reese, M.; McCormick, A. V.; Cussler, E. L. Converting Wind Energy to Ammonia at Lower Pressure. *ACS Sustain. Chem. Eng.* **2018**, *6* (1), 827–834.

(3) Reese, M.; Marquart, C.; Malmali, M.; Wagner, K.; Buchanan, E.; McCormick, A.; Cussler, E. L. Performance of a Small-Scale Haber Process. *Ind. Eng. Chem. Res.* **2016**, *55* (13), 3742–3750.

(4) Cussler, E.; McCormick, A.; Reese, M.; Malmali, M. Ammonia Synthesis at Low Pressure. *J. Vis. Exp.* **2017**, No. 126, e55691–e55691.

(5) Wagner, K.; Malmali, M.; Smith, C.; McCormick, A.; Cussler, E. L.; Zhu, M.; Seaton, N. C. A. Column Absorption for Reproducible Cyclic Separation in Small Scale Ammonia Synthesis. *AIChE J.* **2017**, *63* (7), 3058–3068.

(6) Malmali, M.; Le, G.; Hendrickson, J.; Prince, J.; McCormick, A. V.; Cussler, E. L. Better Absorbents for Ammonia Separation. *ACS Sustain. Chem. Eng.* **2018**, *6* (5).

591297 Material Selection in Process Intensification: Application to Extractive Separation Systems

Abstract:

Extractive separation, a classic process intensification system, offers the solution to overcome the physical equilibrium at azeotropic point by introducing an extra mass separating agent (i.e., entrainer) [1-2]. The material selected plays a key role due to: (i) its solubility properties to enhance the system's mass transfer driving force to drive separation, and (ii) its hazardous properties (e.g., toxicity, flammability) for environmental and safety considerations. Despite recent advances in the synthesis of process intensification systems leveraging phenomenological representation methods to deliver potentially "out-of-the-box" process solutions [3-5], the role of (advanced) materials in process synthesis/intensification has not been sufficiently addressed in open literature [6].
In this work, we propose a systematic approach for material selection, process synthesis, integration, and intensification based on recent extensions of the phenomena-based Generalized Modular Representation Framework (GMF) [7-8]. Herein, processes are represented as aggregated multifunctional mass/heat exchange modules with which intensification possibilities can be discovered without a pre-postulation of equipment or flowsheet configurations. Driving force constraints, derived from total Gibbs free energy change, are employed to characterize mass/heat transfer feasibility by exploiting the general thermodynamic space, and thus result in a more compact modular representation of the chemical systems. Material selection is achieved by incorporating rigorous thermodynamic models (e.g., NRTL) in GMF representation to capture the nonideal liquid mixture properties as well as to assess material performance to facilitate separation. Mass and/or heat integration are also systematically integrated in the superstructure formulation of the modular network to reduce energy consumption, resource utilization, and waste production. The resulting synthesis problem is formulated as a mixed integer nonlinear programming optimization problem (MINLP), where material selection, process synthesis, integration, and intensification can be simultaneously accounted for with respect to process profitability. An extensive case study concerning enthanol/water extractive separation is presented to showcase the proposed approach, with two entrainer candidates examined: a conventional entrainer ethylene glycol (EG) and an ionic liquid entrainer 1-ethyl-3-methyl-imidazolium acetate ([EMIM][OAc]). Extensions of the framework to include process safety and environmental considerations will also be discussed [8]. 1. Tian, Y., Demirel, S. E., Hasan, M. M. F., & Pistikopoulos, E. N (2018). An Overview of Process Systems Engineering Approaches for Process Intensification: State of the Art. Chemical Engineering and Processing: Process Intensification, 133, 160-210. 2. Zhou, T., Song, Z., Zhang, X., Gani, R., & Sundmacher, K. (2019). Optimal Solvent Design for Extractive Distillation Processes: A Multi-objective Optimization based Hierarchical Framework. Industrial & Engineering Chemistry Research. 3. Demirel, S. E., Li, J., & Hasan, M. F. (2017). Systematic process intensification using building blocks. Computers & Chemical Engineering, 105, 2-38. 4. da Cruz, F. E., & Manousiouthakis, V. I. (2016). Process intensification of reactive separator networks through the IDEAS conceptual framework. Computers & Chemical Engineering, 105, 39-55. 5. Tula, A. K., Babi, D. K., Bottlaender, J., Eden, M. R., & Gani, R. (2017). A computer-aided software-tool for sustainable process synthesis-intensification. Computers & Chemical Engineering, 105, 74-95. 6. Stankiewicz, A., & Yan, P. (2019). 110th Anniversary: The Missing Link Unearthed: Materials and Process Intensification. Industrial & Engineering Chemistry Research. 7. Papalexandri, K. P., & Pistikopoulos, E. N. (1996). Generalized modular representation framework for process synthesis. AIChE Journal, 42(4), 1010-1032. 8. Tian, Y., & Pistikopoulos, E. N. (2018). Synthesis of Operable Process Intensification Systems — Steady-State Design with Safety and Operability Considerations. Industrial & Engineering Chemistry Research, 58(15), 6049-6068.

590190 Data-Driven Approach to Systematic Process Intensification Using Building Block Superstructure

Abstract:

Process intensification (PI) involves improvements in process performance that can lead to substantially smaller, cleaner, safer, and more energy-efficient technologies [1, 2]. However, identification of intensified solutions at the conceptual design stage is a challenge as there can be a plethora of possible process configurations. Also known as superstructure optimization, optimization-based process synthesis can be used to identify the optimal process configuration. However, traditional optimization-based approach requires all process configurations and alternatives to be specified beforehand with fixed connectivity and equipment types. Such pre-postulated superstructure may not exist in some cases; thus, the optimal intensified process cannot be identified [3].

A building block superstructure has been proposed recently to overcome the aforementioned challenge [4]. Unlike the conventional superstructure approach, the building block superstructure does not require *a priori* postulation of process alternatives. Instead, each building block is used to represent a fundamental constituent of a unit operation, which can then be combined together to represent physicochemical phenomena. These building blocks can be positioned on a two-dimensional grid, and the optimal intensified flowsheet can be automatically generated by solving a mixed-integer nonlinear programming (MINLP) model [5]. However, as these building blocks are represented using a set of algebraic equations describing complex thermodynamic and material and energy transfer, it contains several non-convex equations, which increases the complexity of the final optimization model. Therefore, locating an optimal solution still remains a challenge [6].

In this work, we propose the use of data-driven approach for the design of a building block superstructure. Specifically, instead of relying on a complicated set of algebraic equations describing physical phenomena, we construct accurate and tractable machine learning-based process models to reduce the complexity of the building block superstructure model [7]. In particular, we use an iterative machine learning-based modeling and optimization algorithm, which involves: 1) constructing an efficient design of experiments, 2) constructing machine learning-based process models, and 3) formulating and solving the machine learning-based MINLP problem. A case study on a reactive separation system with non-ideal thermodynamics will be presented, and known process information and constraints will be considered during the design stage to identify a realistic intensified design.

1. Stankiewicz, A. and J.A. Moulijn, *Process Intensification: Transforming Chemical Engineering*. Vol. 96. 2000. 22-33.
2. Tian, Y., et al., *An overview of process systems engineering approaches for process intensification: State of the art.* Chemical Engineering and Processing - Process Intensification, 2018. **133**: p. 160-210.
3. Demirel, S.E., J. Li, and M.M.F. Hasan, *A General Framework for Process Synthesis, Integration, and Intensification.* Industrial and Engineering Chemistry Research, 2019. **58**(15): p. 5950-5967.
4. Li, J., S.E. Demirel, and M.M.F. Hasan, *Process synthesis using block superstructure with automated flowsheet generation and optimization.* AIChE Journal, 2018. **64**(8): p. 3082-3100.
5. Demirel, S.E., J. Li, and M.M.F. Hasan, *Systematic process intensification using building blocks.* Computers & Chemical Engineering, 2017. **105**: p. 2-38.
6. Boukouvala, F., M.M.F. Hasan, and C.A. Floudas, *Global optimization of general constrained*

grey-box models: new method and its application to constrained PDEs for pressure swing adsorption. Journal of Global Optimization, 2017. **67**(1): p. 3-42.
7. Kim, S.H. and F. Boukouvala, *Machine learning-based surrogate modeling for data-driven optimization: a comparison of subset selection for regression techniques.* Optimization Letters, 2019.

589503 Dow LP Oxo ^{SM} Variable Selectivity Technology - Greater Than the Sum of Its Parts

Abstract:

Hydroformylation – the conversion of C3 and higher olefins to a mixture of linear and branched aldehydes- is a widely practiced application of homogeneous transition metal catalysis. While the linear product is typically preferred, shifts in market demands or product portfolios may at times favor the branched isomer; thus the ability to alter the product ratio constitutes an operational advantage. Dow LP Oxo ^{SM} Variable Selectivity Technology (VST) featuring a rhodium catalyst promoted by tandem ligands, may be used to produce aldehyde mixtures comprising a broad range of isomer ratios. The relative propensities of the two ligands for forming rhodium complexes and the varying kinetics of the resulting active species allows impressive control of product regioselectivity through a number of process variables. The underlying principles and associated examples of VST will be discussed.

LP Oxo is a service mark of The Dow Chemical Company ('Dow') or an affiliate of Dow.

587855 Utilizing Failure Mode and Effects Analysis (FMEA) to Improve Commercialization Process

Abstract:

Demands for new products with high quality and increased speed to market continue as resources remain limited. Two risk management tools, Design Failure Mode and Effects Analysis and Process Failure Mode and Effects Analysis (FMEA), are often completed in the design phase of commercializing new products due to requirements from a customer or as part of a corporate project execution strategy. Changing the mindset and discussion around the use of these tools from a project execution check list to a source of valuable information can improve quality, while also reducing costs and improving speed to launch for these new products.

A case study will be used to demonstrate the value these tools can bring to commercialization. A Design FMEA was used to identify key areas of the product formulation that would lead to customer failures. A set of experiments was completed based on results from the FMEA scoring that exposed an unexpected failure mode. Further investigation revealed a product re-formulation was required, the experimentation driven by the FMEA output played a key role in defining the new formulation. A Process FMEA was then created that identified crucial process steps that were studied at a small scale. Information gained allowed for a successful initial scale-up improving time to market for this product and reducing costs associated with the scale-up activity.

Overall the systematic team approach to creating FMEA documentation throughout a project can lead to improvements in work planning and execution. Such methods reduce time to commercialization improving resource efficiency while enhancing the likelihood of higher market success.

591511 Electrification, Process Intensification, and Artificial Intelligence for Modular and Improved Chemical Manufacturing

Abstract:

There is renewed interest in the chemistry and processing of small molecules, such as carbon dioxide, CO_2, ammonia, NH_3, and small hydrocarbons (methane, ethane and propane). The large energy demand and associated large production of CO_2 in ammonia production, the conversion of CO_2 to chemicals and fuels, and the revolution of shale gas in the U.S., and thus of small hydrocarbons, are changing the global energy and chemicals terrain and are creating new opportunities for research and manufacturing. Common to the small molecules are the challenges of activating chemical bonds, the large energy demand, the equilibrium limitations, the loss of selectivity, and catalyst deactivation. In addition, processing some of these resources may require distributed manufacturing, whose economic viability is uncertain. These challenges require understanding of the fundamental chemistry and development of new catalysts and processes. This talk will overview these emerging trends and present examples from chemical activation using electrification, such as microwaves, coupled with other process intensification concepts to allow scalability. The role of data science in process design will also be discussed.

590197 Advanced Integrated Techniques for Quantifying Kinetics and Transport Parameters in Multiphase Reactors

Abstract:

Multiphase reactors and multiphase flow systems have found extensive related to energy, environmental, chemicals, petroleum and variety of products processes. Multiphase flow reactors are complex due to the intricate interactions and reactions among difference phases. Therefore, sophisticated techniques that integrate measurements in a novel way of various hydrodynamics and transport parameters and their integration with kinetics are required to quantify and visualize these systems to provide benchmarking data for validating CFD and models. These advances should provide the mechanistic frame work to properly scale up the results from lab to commercial scales.

In our laboratory, various types of multiphase systems such as slurry/basket reactors and packed bed reactors under high temperature and high pressure for various industrial and environmental processes have been studied by developing advanced measurement techniques and validating CFD simulations which enable process intensification and facilitating the implementation of new mechanistic scale-up methodologies. In general, these techniques can be classified into two categories: radioisotope techniques and non-radioisotopes techniques. The radioisotope techniques include radioactive particle tracking (RPT), dual source gamma ray tomography (DSCT), gamma ray densitometry (GRD) for 3D flow field, velocity and turbulent parameters, phases distribution and flow pattern identification measurements. The non-radioisotope techniques include 4-point optical probe for bubble dynamics, optical probes for gas-solid dynamics, optical probe for local mass transfer, optical probe for local liquid and gas velocities, holdup and their time series distribution in packed beds, gas tracer technique for global mass transfer, heat transfer probe, pressure transducers, and others.

591298 Process Intensified Conversion of Stranded Energy Resources to Value-Added Liquid Products for Distributed Production

Abstract:

Over 200 billion ft^3/year of natural gas at remote production sites is flared in the U.S. due to the limitations in pipeline transportation capacity. This wastes valuable resources, increases air emissions, and burns profits. Similarly, renewable electricity (wind and solar) in some disadvantaged locations becomes stranded due to the limitations in power transmission capacity and intermittent nature of the renewable electivity. This presentation introduces the applications of microwave catalysis for the conversion of these stranded energy resources into liquid transportable chemicals such as aromatics and ammonia. Microwave irradiation has shown a profound impact on catalyzed and uncatalyzed gas-solid reactions. In particular, it has been demonstrated that microwave-specific effects can manifest themselves through the enhancement of reaction rates, changes in the position of equilibria and the distribution of products.

Methane must be activated to a high-energy transition state before it can further react to produce stable products. The high reaction temperatures (>700°C) required to activate methane's strong C-H bonds usually results in rapid and extensive coke formation that blocks catalyst micropores, deactivating the catalyst. The gaps in scientific understanding of the direct, non-oxidative methane conversion process preclude the design of highly selective catalysts and reactors to make desired higher value products. Technical hurdles include catalyst stability and thermodynamic limitations that constrain the desired product yields. In this study, methane conversion reaction is enhanced with microwave irradiation to demonstrate microwave reaction chemistry at catalyst-reactant interface for selective C-H bond activation

Stranded electricity can be converted to liquid ammonia under ambient pressure and 320°C using water and air as feedstock. Conventional ammonia synthesis process, i.e. Haber-Bosch process, cannot be scaled down economically due to the high temperature and high pressure operation and the intermittent nature of renewable electricity that requires frequent startup and shutdown. Catalytic ammonia synthesis was conducted under two microwave irradiation scenarios: microwave and microwave plasma. Other than selective activation of dinitrogen to metastable radicals, the most obvious advantage that microwave irradiation affords in driving a heterogeneously catalyzed reaction is the ability to locally heat the catalytic sites. Many industrial processes utilizing heterogeneous catalysts are high-temperature processes wherein both components of the reaction (i.e., catalyst and medium) are heated to the temperature required for the reaction to occur. Specifically discussed in this presentation are the effects of electromagnetic properties of catalysts, microwave frequency and microwave energy absorbed on the conversion of nitrogen and ammonia yield.

Technoeconomic analysis was carried out to compare microwave catalytic processes with the state-of-the art. The comparisons are made to focus on process intensification elements such as energy efficiency, energy productivity and capital cost reduction. Sensitivity analysis was made to evaluate various scenario and different plant sizes. The resulting engineering innovation includes plant siting flexibility due to smaller scale, lower capital and operating costs, higher energy efficiency, lower carbon footprint, no or low environmental emissions.

The potential for a dramatic transformation in the chemical process industries is emerging from the rapid development of process intensification. This can potentially yield substantial economic and environmental benefits in combination with advanced manufacturing by significantly reducing the size of chemical process plants. This presentation aims to revolutionize the chemical process industries by enabling the development of intensified and, where appropriate, modular process plants.

590682 GT-BTX Plus®: Minimize Octane Loss While Maximizing Petrochemicals and Jet Fuel/Diesel Production from FCC Gasoline

Abstract:

GT-BTX PluS®: Minimize Octane Loss While Maximizing Petrochemicals and Jet Fuel/Diesel Production from FCC Gasoline

GT-BTX PluS® is a unique and industrially proven extractive distillation-based technology that positions a refiner to meet stringent gasoline sulfur requirements with minimal octane loss, Additionally, it capitalizes on increased demand for petrochemicals, and maximizes Jet Fuel/Diesel margins under the MARPOL scenario.

FCC gasoline contributes approximately 30-50% to a refinery's gasoline pool and is the primary contributor of sulfur. Implementation of stringent gasoline sulfur requirements has led to increased octane loss across the FCC gasoline block due to a loss of valuable high-octane olefins. GT- BTX PluS® minimizes this octane loss to <0.5 RON points by preservation of high-octane olefins, while meeting sulfur requirements.

The implementation of GT-BTX PluS® on FCC naphtha results in two different petrochemical precursor streams: aromatic rich extract and olefinic rich raffinate. By proper routing and processing of these streams in different units, GT- BTX PluS® enables the refiner to produce olefins and aromatics from FCC gasoline. This enables the FCC based refiner to transition from a fuels-based platform into a more profitable petrochemicals producer.

With the implementation of IMO 2020, there is speculation that the replacement of bunker fuel with MGO/VGO can result in increased Jet Fuel/ diesel margins up to USD 20/bbl. Further, it has been projected that an increase in air travel in developing countries will result in jet fuel being the fastest growing transportation fuel until 2050. By proper optimization of FCC pre-treat and FCC operations coupled with GT- BTX PluS®, the FCC refiner can boost its LCO production without experiencing higher octane loss in FCC gasoline post treat, resulting in more Jet Fuel/Diesel production. Another derived benefit will be the flexibility to shift from maximum gasoline mode to maximum diesel mode. Conventionally, this shifting results in possible salt deposition and operational issues in the FCC Main Fractionator. GT-BTX PluS® offers the unique advantage of continuing to run in the max gasoline mode while still reaping the benefits of higher Jet Fuel/Diesel production.

591336 Process Intensification of Ternary Distillation Using Dynamic Optimization Method and Data-Driven Approach

Abstract:

Distillation is the dominant separation technology for liquid mixtures in chemical industry. Studies revealed that over 40,000 distillation columns in operation in the US are responsible for more than 90% of purification needs [1]. The thermodynamic efficiency of distillation can be as low as 10% [2], and the operation of distillation can account for half of the energy consumed by chemical plants [3].

Process intensification (PI) is by now an accepted means for improving the energy efficiency of distillation column (sequences). Through the implementation of innovative design changes, process intensification significantly lowers both capital and operating costs. However, tight design integration/integration comes with control challenges: a reduced number of control degrees of freedom and a more constrained operation window compared to conventional columns [4].

In this paper, we rely on a different approach to process intensification, implemented in the temporal domain, which we refer to as "dynamic process intensification (DPI)." DPI aims to improve process efficiency by exploiting and favorably manipulating in the time domain the nonlinear dynamics of existing columns [5-7]. The principle of DPI consists of imposing operational instead of design changes, aiming to meet product quality constraints while minimizing energy per production rate through implementing a transient (periodic) operating pattern. Thus far, such patterns were defined in terms of switching between a finite (generally, two) fixed operating points. In this paper, we introduce a novel approach based on dynamic optimization. We introduce a data-driven approach for learning the DPI optimization-relevant dynamics of a distillation process from its operational history and present a dynamic optimization-based DPI framework. We show that, among others, Hammerstein-Wiener (HW) models that have fewer parameters to regress and can be trained with a considerably smaller amount of data can accurately capture the dynamics of the distillation plant along with its control system [8-9].

We will apply this conceptual framework to an extensive case study separating cyclohexane, toluene and m-xylene mixture. Meeting all quality constraints, DPI can offer considerable energy savings compared to steady-state operation of the same column.

References

[1] Humphrey, J. L. Separation processes: playing a critical role. *Chemical Engineering Progress,* **1995**, 91.na

[2] Humphrey, J. L.; Seibert, A.; Koort, R. Separation Technologies: Advances and Priorities,

tech. rep., Office of Industrial Technologies, Washington DC, **1991**.

[3] Isopescu, R.; Woinaroschy, A.; Draghiciu, L. Energy reduction in a divided wall distillation column. *Revista de Chimie* **2008**, *59*, 812-815.

[4] Schembecker, G.; Tlatlik, S. Process Synthesis for Reactive Separations, *Chem. Eng. Process*, **2003**, *42*, 179-189.

[5] Baldea, M.; Edgar, T. F. Dynamic Process Intensification, Dynamic Process Intensification, *Current Opinion in Chemical Engineering*, **2018**, *22*, 48-53.

[6] Yan, L.; Edgar, T. F.; Baldea, M. Dynamic Process Intensification of Binary Distillation Based on Output Multiplicity, *AIChE J*, **2019**, *65(4),* 1162-1172.

[7] Yan, L.; Edgar, T.; Baldea, M. Dynamic Process Intensification of Binary Distillation via Periodic Operation, *Industrial & Engineering Chemistry Research,* **2019**, *58(15)*, 5830-5837.

[8] Richard C Pattison, Cara R Touretzky, Ted Johansson, Iiro Harjunkoski, and Michael Baldea. Optimal process operations in fast-changing electricity markets: framework for scheduling with low-order dynamic models and an air separation application. *Industrial & Engineering Chemistry Research*, 55(16):4562–4584, 2016

[9] Ian Goodfellow, Yoshua Bengio, and Aaron Courville. Deep learning. *MIT Press*, Cambridge, MA, 2016.

589766 Process Intensification of CO_2 Absorption Using an Additively Manufactured Intensified Packing Device

Abstract:

A multifunctional packing device has been developed, characterized, and tested for process intensification of CO_2 absorption by aqueous amines in packed columns. The multifunctional device is an additively manufactured packing element that integrates effective contact of phases and heat exchange in multiphase systems, and can be employed to vent excess heat from an exothermic reactive system or supply thermal energy to an endothermic chemical reaction. The impact of excess heat removal by the device on the absorption of CO_2 by aqueous solutions of monoethanolamine is investigated. The design process of the device and the validation of core hydrodynamic metrics will be discussed. Heat and mass transfer testing demonstrated that the device was effective at removing excess heat from the reaction and, furthermore, that cooling could successfully enhance capture of CO_2 by the amine solvent. The magnitude of the enhancement was found to depend strongly on operating conditions, in particular flow rates of CO_2, air, and solvent concentration. This device represents a successful example of the role of additive manufacturing in the intensification of chemical processes.

590385 An Efficient and Reliable Numerical Method for Computing Coupled Multicomponent Fluxes for Complex Mixtures across Asymmetric Membranes

Abstract:

An Efficient and Reliable Numerical Method for Computing Coupled Multicomponent Fluxes for Complex Mixtures Across Asymmetric Membranes

Dylan Weber[1], Ronita Mathias[1], Ryan Lively[1], Chau-Chyun Chen[2], Joseph Scott[1†]

1 Department of Chemical and Biomolecular Engineering, Georgia Institute of Technology, Atlanta, GA 30332, USA

2 Department of Chemical Engineering, Texas Tech University, Lubbock, TX, 79409, USA.

Corresponding author; Email: joseph.scott@chbe.gatech.edu

Abstract

Membranes offer a promising means to intensify many industrial separation processes that are currently achieved using distillation and other thermal separation technologies. Historically, key limitations in the design and synthesis of high-performance membrane materials has posed the most significant barrier to the adoption of membrane processes in industry. However, with steady improvements in membrane material properties and module construction, the lack of reliable membrane modeling and simulation tools is emerging as a major barrier in its own right. In particular, widespread industrial adoption of membrane processes as an alternative to conventional thermal separations is unlikely to occur in the absence of (i) reliable predictive models for membrane performance when confronted with complex multicomponent streams, (ii) methods and databases for parameterizing such models using pure component data and limited multicomponent measurements, and (iii) efficient and reliable numerical solution methods that are compatible with standard process flowsheeting software and enable technoeconomic comparison of membrane and non-membrane processes on a consistent basis.

In this presentation, we will describe a novel numerical solution procedure for computing transmembrane fluxes for non-ideal multicomponent mixtures in contact with an asymmetric membrane. Numerous relevant applications of membrane separations such as water purification, carbon capture, hydrogen separation, olefin/paraffin separation, and benzene derivative concentration include process streams that are multicomponent nonideal mixtures. Along with that there are many different types of membrane material being used. From highly selective zeolites, carbon nanotubes, ionic polymers, mixed-matrix MOF and polymeric, graphene-oxide, and specialty polymers. Many of the current approaches for industrial practice adopt the permeability and linear driving force model to describe permeation across the selective layer. These models are applicable in many cases where the sorption is ideal and uncoupled transport can be assumed. However, this explanation of the system when trying to model many nonideal components and interacting permeants is not correct. That is why our group is looking to employ a feasible rigorous solution of the Maxwell-Stefan equations to describe the multicomponent mass transport. Past contributions from experts in these equations such as Rajamani Krishna, Benny Freeman, or Pavel Izák employ approximate solutions by evaluating terms at average compositions, assume the permeate composition is zero to allow for an approximate solution, or use finite differences to approximate the differential terms. These studies are also restricted to a ternary system with two permeates and a membrane phase. No work has been done to have an effective and reliable numerical method for a system with any number of components.

As a brief overview, the Maxwell-Stefan (MS) approach towards mass-transport is based on irreversible thermodynamics by balancing the driving forces any component from a mixture experienced in the form of the chemical potential with the collective friction forces experienced by transporting with other components in the mixture. The Maxwell-Stefan diffusivities can be physically interpreted as the inverse friction or drag coefficients of a binary interaction between two permeating molecules or the permeant and the membrane phase. The main advantages of this transport model compared to the Fickian model is the incorporation of driving forces other than a concentration gradient, but namely MS diffusivities are not dependent on the system and are an intrinsic property of the binary component or membrane pair. Note, these diffusivities can still be functions of temperature, pressure, and composition. The difference is the general applicability from single or binary permeation data to the multicomponent system with many nonidealities accounted for. Then with the nature of many nonidealities, there are many problems in solving such systems of equations. The problems associated with the complete solution of the set of MS equations are the numerical methods associated with solving the matrix formulation of the set of coupled ordinary differential equations. Our research goal is to provide a working rigorous solution method that converges to physically meaningful solutions with only feed stream inputs, a thermodynamic activity model with required parameters, and single component membrane diffusivities (in some cases binary component diffusivities). This will provide a working toolset for the practicing chemical engineer to design these systems without having to be an expert in the underlying transport equations and allow for rapid plant design prototyping and screening of various process units. In this work we consider the Maxwell-Stefan as the general transport model, and the multicomponent Flory-Huggins equation to describe the activity of the permeants in the polymer phase. The interesting note is this contribution is applicable to any membrane system that can be described by the Maxwell-Stefan equations.

590118 Ethylene Production Using Oxidative Dehydrogenation: Effects of Alternative Separation Technologies on Process Safety & Economics

Abstract:

Summary

Previous work from this group utilized a HYSYS simulator to predict the performance of an ethylene oxidative dehydrogenation (ODH) production facility using alternative diluents and separation technology not currently used in the standard industrial process [*1-3*]. An ODH process system has been modeled with ethane being fed along with air into a reactor with simultaneous mixing with pure methane to lower the product gas temperature. The methane feed was later separated from the other product gases and eventually fed into a recycle stream. This process was fairly expensive due to the use of a demethanizer within the process recycle stream. Safety and operational improvements for the ODH process was another aspect of this analysis that yielded interesting results. By the inclusion of storage vessels before and after the ODH reactor, the reactor unit can operate even in the event of an upstream or downstream failure, resulting in less downtime for the process.

The current work investigated the development of alternative separation technologies for the production of ethylene for the oxidative dehydrogenation (ODH) reaction of ethane using a HYSYS simulation with economic and safety analyses. The application of polymer-membrane separation and non-cryogenic distillation alternatives, such as bottom-flash, heat-pump assisted and low-temperature distillation, were simulated and evaluated individually and in-tandem with membrane separation. The effects of alternative diluents, process flammability issues, and economic analysis results will also be discussed.

Literature Cited:

1.Maffia GJ, Gaffney AM, Mason OM, Techno-economic analysis of oxidative dehydrogenation options, Top. Catal. (2016), http://dx.doi.org/10.1007/ s11244-016-0677-9 (2016, August 16).

2.Baroi C, Gaffney AM, Fushimi R, Process economics and safety considerations for the oxidative dehydrogenation of ethane using M1 catalyst, Catal. Today (2017), http://dx.doi.org/10.1016/j.cattod.2017.05.041 R (2017, May 11).

3.Gao, Y., Neal, L., Ding, D., Wu, W., Baroi, C., Gaffney, A.M., Li, F., Recent Advances in Intensified Ethylene Production: A Review, ACS Catalysis (2019), http://10.1021/acscatal.9b02922 (2019, August 12).

614862 Energy Efficient Separations of Olefins and Paraffins through a membrane II

Abstract:

Update on the Separation of Olefin-Paraffin Mixtures with Optiperm^{TM} Customized Amorphous Fluoropolymer Membranes

Abstract: The separation of light olefins from paraffins is one of the most energy intensive in the world. The state of the art separation method is distillation, a technology that has remained unchanged over decades. Membranes hold the promise of a more modular, energy efficient and capital efficient method for separation. However, they have historically been hampered by low performance and stability.

Compact Membrane Systems has developed a novel membrane material using a silver embedded fluoropolymer material to transport olefins across the membrane. This material has demonstrated over 2 years of longevity in laboratory operation and has shown high selectivity and permeance under realistic operating conditions. Additionally Optiperm membranes were trialed at the Delaware City Refining Company in over 300 days of field operations validating the robustness and capability of the technology in an operating plant environment.

During the past year, CMS has developed and installed a laboratory recycle test systems to allow for the continuous testing of spiral wound modules with areas of ~20 square feet. The recycle test unit enables CMS to test larger scale membrane modules under field operating conditions using actual process streams from end users. Furthermore the recycle rig design incorporates our innovative design for hydrating the process feed within the membrane housing, significantly reducing the complexity of the membrane systems operations.

We will share the latest leanings from our recycle rig testing as well as provide updates on the scale up of our flat sheet spiral wound membrane cartridges. We will also discuss the current state of our hollow fiber membrane configuration development and the progress on our joint project with Dow Chemical through RAPID, a manufacturing institute funded by the Department of Energy. The goal of the work with Dow is to produce a chemical grade propylene from a process feed stream with a propylene concentration of greater than 60%.

CMS will also discuss results on laboratory testing of C_4 streams. This includes mixed gas separation of n-butane and 1-butene with permeances between 200-500 GPU and selectivity greater than 20. We will share results for a number of operating conditions and will highlight several exciting C_4 separation applications including those with isomer streams. In addition, longevity results will be shared demonstrating the stability of the membrane in C_4 separations.

591159 When Disasters Happen: A Collaborative Approach Towards the Community's Recovery Process

Abstract:

The 2020 Virtual AIChE Spring Meeting & 16th GCPS Wednesday keynote "When Disasters Happen: A Collaborative Approach Towards the Community's Recovery Process" will be presented by Thomas Munoz, Emergency Manager and Homeland Security Director with Texas City. As the Emergency Manager and Homeland Security Director with Texas City, the home of 19 chemical plants, he is responsible for the safety of all its citizens along with managing the crisis communication plan with local industry and major stakeholders.

Safety and disaster planning is vital to the recovery process after a major incident. As key critical infrastructure industries, issues may arise that have not been planned for in advance, such as the recovery from an incident. Developing and maintaining partnerships with local partners in your community will support addressing unforeseen challenges. Recovery efforts include planning with key partners and understanding the phases of recovery, which also consider the mental health and well of everyone involved in the response, as well as the community impacted. In this keynote address a discussion of pre-incident planning considerations and strategies as well as lessons learned/after action improvements post-recovery will be discussed. Additionally, collaboration considerations regarding local resources to promote recovery in the event of a workplace incident will be discussed.

C0-presenter Amanda Groller, MS.Ed., LPC will discuss the mental health impacts of a disaster.

589942 Successes and Failures of Pilot Plants for Technology Commercialization in Chemicals and Fuels: A Look Back over 3 Decades

Abstract:

Pilot plants for commercial deployments in Shell over the past three decades have included fully integrated units, and virtual integration using modeling to integrate piloting of select unit operations, as well as data from bench- scale units and cold-flow experiments. The successes and issues for these pilot plants and demonstration units will be reviewed, including challenges in scale-down and solids handling which impact trade-offs between risk-reduction accuracy, timing, and development cost. Adaptations of existing pilots to new or modified projects and use of pilot plants and bench scale units to address discrepancies or problems learned after commercial startup will also be presented for assessment of best practices and learnings for the use of pilot plants in process development and technology commercialization.

589015 Lessons Learned in Use of Small-Scale Distillation Columns

Abstract:

Success in piloting new process developments involves integrating process models, technology fundamentals, and defining critical operating parameters for the experimental plan. While it is important that test plans are developed to address key parameters, understanding the limitations of the pilot unit helps to identify how the obtained data can be applied for final industrial scale design. The results of the studies presented here provided insight into using small pilot distillation columns to validate process models and provided important design considerations for commercial scale operation. Key learnings in operating small scale pilot distillation columns and applying data for design considerations is the focus of this presentation using a propylene oxide purification method as an example.

OPTIMAL DESIGN OF A LIQUID-LIQUID EXTRACTION SYSTEM

Authors:

Donald J. Glatz
Brendan Cross

Koch Modular Process Systems, LLC

Prepared for Presentation at

2020 AIChE Spring Meeting
August 2020

INTRODUCTION

In most separation applications involving liquid-liquid extraction (LLE), the process will require one or more distillation columns downstream of the extraction column. Thus, optimization of the process requires a focus on the overall integration of extraction and distillation columns in order to provide the most economical solution. Typically, the solvent used for the extraction step must be distilled from the extracted solute and then recycled back to the extraction column. Also, if the solvent has some solubility in the feed, as is often the case for systems that include an aqueous feed and an organic solvent, then a distillation column will be required to recover and recycle the solvent from the aqueous phase (raffinate). As discussed in previous papers [1] [2], design of the extraction column almost always requires pilot plant testing in order to provide data for accurate scale-up and design of the commercial column. Often the distillation columns can be designed by first simulating the process and then designing the distillation column internals using known design criteria, hydraulic rating software and process engineering experience. However, in some instances, such as foaming, and unknown interactions between various components in the system, pilot plant testing of downstream distillation steps may also be required in order to provide a firm basis for the commercial process design. This paper will present three case studies involving the design of complete LLE systems with downstream distillation. In each case, the extraction column was sized via pilot plant testing. In two of the applications, the distillation columns were designed via simulations only, and in the other application a distillation column was designed via pilot plant testing plus simulation. All three system have been successfully operating in the commercial installations.

Case Study 1 – Extraction of Acetic Acid from Fermentation Broth

This application involved the extraction of acetic acid from an aqueous fermentation broth generated in a cellulosic ethanol process. Ethyl acetate (EA) was the selected solvent. The overall process for this application was expected to be as shown in Figure 1. The first column is an extraction column where EA is used to counter currently contact the broth and extract the acetic acid. The extract phase (EA) leaving the top of the column is directed to a distillation column (the solvent recovery column) to regenerate EA for recycle back to the extraction column, and to remove and recover the acetic acid that was extracted from the EA. Aqueous raffinate leaving the bottom of the extraction column will contain the soluble amount of EA and thus must be sent to a stripping column to recover the EA and also to recycle this solvent back to the extraction column. As shown in Figure 1, the overheads from each distillation column enter a common decanter which produces an organic, EA rich phase and a heavier aqueous rich phase. A portion of the EA rich phase is recycled back to the extraction column, with the balance going back to the solvent recovery column as reflux. The aqueous rich phase is sent back to the stripping column as reflux. As in most 3-column extraction process such as this, the majority of the operating cost for the system is in the extract distillation column. Thus, operating cost is minimized by optimizing the process such that the amount of EA used in the extraction step is minimized and the distillation column is designed to operate in a manner that minimizes the utilities required. Using less EA in the extraction step also minimizes the capitol cost of the solvent recovery column.

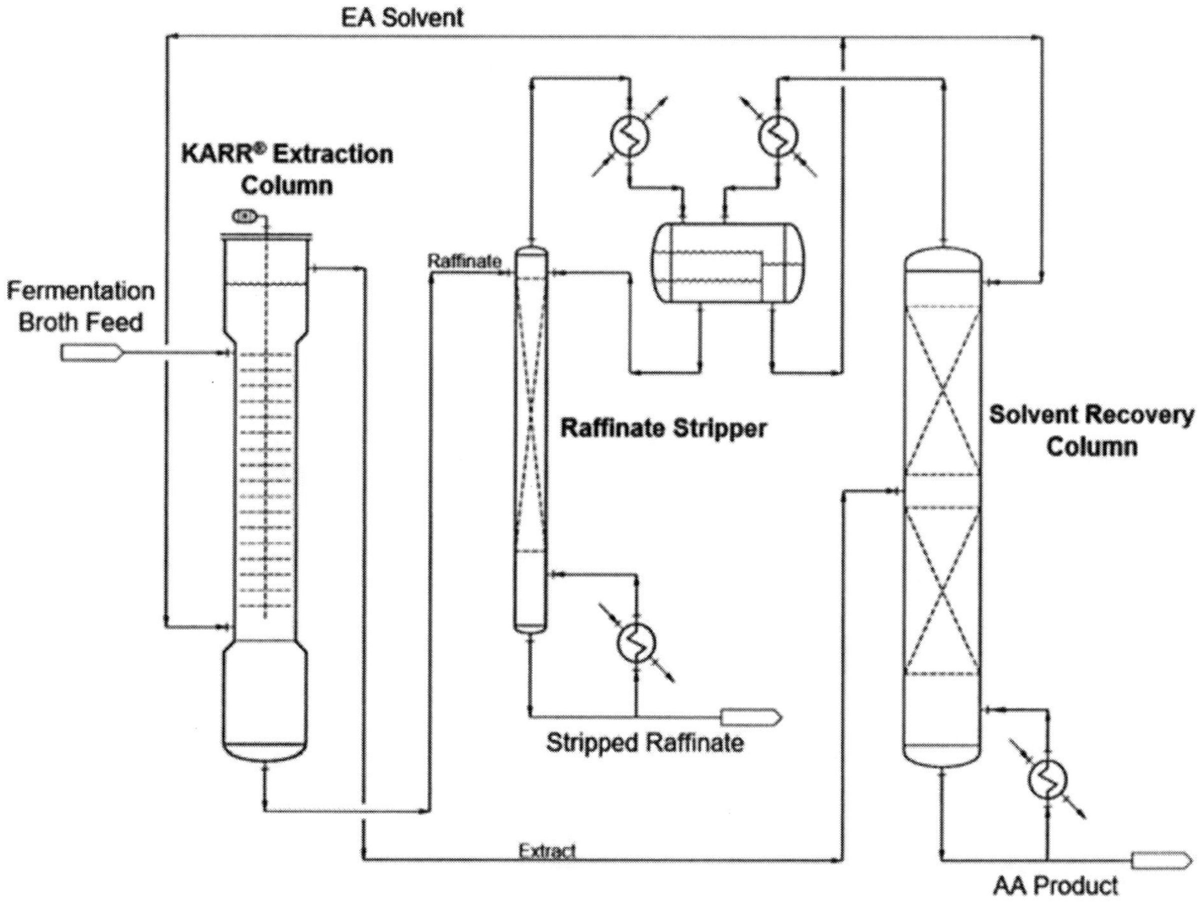

Figure 1 Extraction of Acetic Acid from Fermentation Broth Flow Diagram

Initial investigation into the parameters for this process indicated that sufficient vapor-liquid equilibrium (VLE) data was available for the components in the system. Also, the process engineers involved had the design experience and the appropriate modeling software in order to effectively simulate and design the distillation and stripping columns. The key design information required was the material balance around the extraction column. Selection of the appropriate extraction column technology and pilot plant testing was required in order to optimize the performance of the column and develop the corresponding material balance.

Optimization of the extraction process has been reported in a previous paper [3]. Due to a strong tendency toward emulsification, the KARR® Column was selected for this process [4]. Results from pilot plant testing generated the following design criteria:

- Scale-up was based upon a 1" diameter x 12' plate stack height, pilot KARR® Column with metal perforated plates. The column was tested with the solvent phase dispered.
- The solvent to feed ratio was 1.5 on an overall, weight basis. I.e. the flow rate of EA was 1.5 times the flow rate of the fermentation broth in order to achieve 98-99% acetic acid extraction.
- The design temperature was 40-45 °C.

- The design capacity was 650 GPH/ft^2 (combined flow of feed + solvent divided by the cross-sectional area of the column).

The material balance developed around the KARR® Column, as a result of the pilot test, for the commercial scale system is shown in Table 1. Based upon this material balance, simulations were performed and optimized for the process design of the distillation and stripping columns and then the mechanical designs for these columns were performed. Thus, the process flow diagram shown in Figure 1 is essentially the same as the process flow diagram that was created for the industrial scale acetic acid extraction process. A complete modular system [5] containing all of the extraction and distillation columns, tanks, heat exchangers, pumps, piping, instrumentation and controls was designed and fabricated off site and delivered to the end user.

	Feed	**Solvent**	**Raffinate**	**Extract**
$\dot{m}$ (g/min)	90	135	87	134
x_{AA}	0.052	0.0010	0.0029	0.035
x_{EA}	0	1.0	0.081	0.93
x_{H2O}	0.95	0	0.92	0.040

Table 1 Extraction of Acetic Acid from Fermentation Broth – Pilot Test Material Balance

Case Study 2 – Extraction of Acids from Wastewater

The application involved the extraction of a mixture of short chain organic acids from a wastewater stream using an alkane solvent which has essentially no solubility in the water, and thus a raffinate stripping column would not be required. The process is shown schematically below in Figure 2. The first column is an extraction column where the alkane is used to counter currently contact the aqueous feed containing approximately 4% acids and extract these into the solvent phase. The desired raffinate acid concentration was < 800 ppm, however from the standpoint of a process performance guarantee, <1000 ppm would be acceptable. The extract phase (solvent) leaving the top of the column is directed to a distillation column to recover the solvent overhead for recycle back to the extraction column. Aqueous raffinate leaving the bottom of the extraction column will have a sufficiently low level of solvent that it can be discharged as treated wastewater from the system.

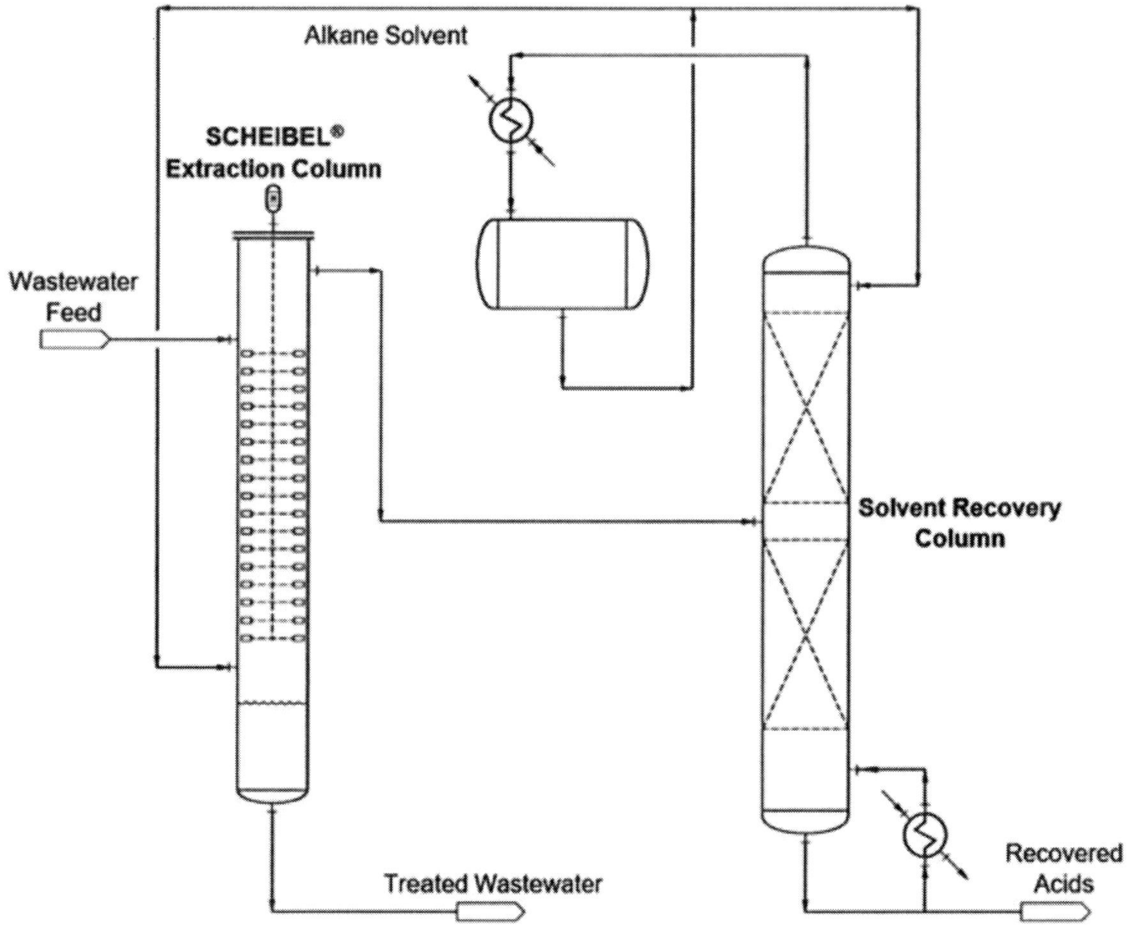

Figure 2 Extraction of Acids from Wastewater Flow Diagram

Both liquid-liquid equilibrium data (for water – organic acids – alkane) and VLE data for the organic acids/alkane separation were obtained from the end user based upon testing in their specified laboratory. The liquid-liquid equilibrium data was confirmed via additional laboratory tests using the actual feed to be fairly accurate. Using the VLE data provided, simulations demonstrated that the overhead solvent could be distilled to achieve very low ppm levels of acid in the recycled solvent. As in Case Study 1, minimizing the amount of alkane solvent used in the extraction column would result in the most economical operating cost for the process in addition to reducing the capitol cost for the downstream solvent recovery column.

Due to a high difference in density between the two phases (approximately 0.3 difference in the specific gravities) and the expectation that the interfacial tension would also be high, a 3" diameter x 60-stage SCHEIBEL® Column [4] was selected to pilot test the extraction step. Initial testing using fresh solvent (no acid) demonstrated successful performance of the column with scale-up [6] capacities up to 600 GPH/ft^2. The minimum solvent to feed ratio that could achieve a raffinate acid concentration below 1000 ppm was 3.5 on an overall, weight basis.

In order to confirm the VLE data received for extract phase separation, pilot plant distillation testing was conducted. The distillation was performed in a 4" diameter x 30' glass column. The column contained 21 feet of 316SS Koch Glitsch BX wire gauze structured packing. There were numerous potential locations for the feed point such that the heights of both the stripping section and the rectification section could be varied. The column was traced with electric heat tape to overcome any heat losses, and then insulated.

Pilot plant testing of the solvent recovery column was unable to achieve the results expected via simulation. It was apparent that the vapor-liquid equilibrium data provided did not reflect the true nature of the separation at the purity required. The test program demonstrated that the distillation column could produce a recovered solvent distillate at several reflux ratios with an acid content between 50-100 ppm acid in the alkane distillate, a level that was significantly higher than simulations predicted.

Based upon these results, additional extraction tests were performed in the pilot SCHEIBEL® Column using solvent spiked with 100 ppm acid. These tests demonstrated that the desired raffinate concentration of < 1000 ppm acid could not be achieved. The concentration of acid leveled out at just above 1000 ppm in the raffinate for testing with S/F ratios between 3.5 and 5.0. Thus, there is a pinch point at this concentration of acid in the solvent. As such a lower level of acid in the recycled solvent is required for this process to achieve < 1000 ppm in the raffinate. Additional tests were then performed using 50 ppm acid in the solvent. The performance guarantee level of acid in the raffinate was achieved at S/F = 1.5, but a ratio of 3.5 was required to achieved less than the desired concentration of 800 ppm.

Like Case Study 1, the process was modeled using a simulation software based upon the pilot test results, and the distillation column was designed using distillation internals hydraulic rating software. A complete modular system was designed and fabricated. The full system was eventually installed and started up with excellent results. A modular system including the SCHEIBEL® Column for extraction and a distillation column with structured packing was fabricated. Since a recycle solvent with 50 ppm acid could not be achieved in the test distillates, and upon agreement with the end user, the design was based upon a raffinate guarantee of <1100 ppm acid and a S/F of 3.5. This system has successfully operated for a number of years. Future plans were to further optimize the process in order to reduce the solvent to feed ratio and/or to achieve 800 ppm or lower levels of acid in the aqueous effluent.

Case Study 3 – Phenol Extraction System

Prior to working on this project, the engineers involved had previous experience with the design, supply and operation of phenol extraction systems. These systems involve the typical, 3-column design as follows:

- An extraction column using a low boiling solvent to extract phenol
- A distillation column for recovery of the phenol (bottoms stream) and recovery of the low boiling overhead solvent.

- A stripping column to strip soluble solvent out of the aqueous raffinate steam, recover this solvent and recycle it to the extraction column, generating wastewater with <1 ppm solvent.

Therefore, the flow diagram for this system as shown in Figure 3 below is analogous to Figure 1 for Case Study 1.

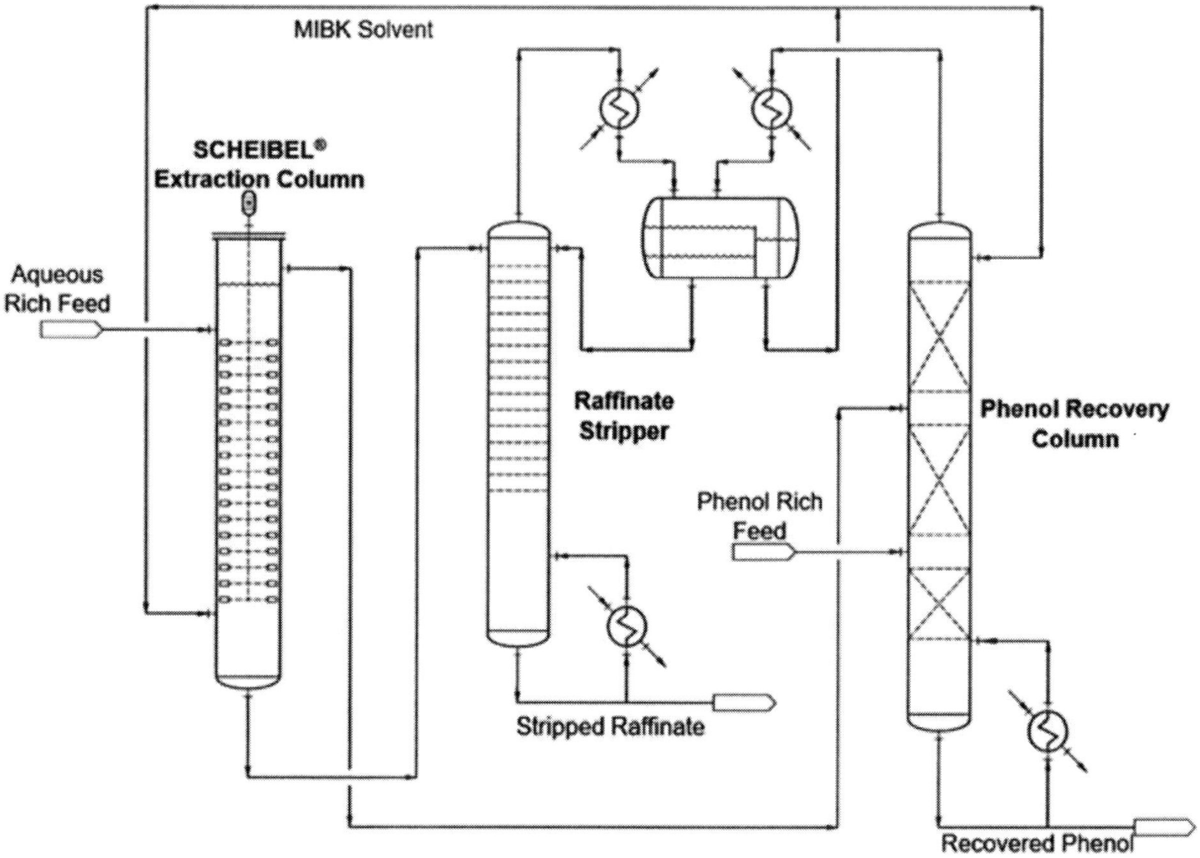

Figure 3 Phenol Extraction System Flow Diagramp

Most phenol extraction systems utilize methyl isobutyl ketone (MIBK) as the solvent due to a very high distribution coefficient for aqueous steams with pH below 7. The design of the distillation and stripping columns are relatively straight forwards via simulations and scale up. However, the design of extraction column can be significantly affected by trace components that are in the feed. As such, in order to accurately design the extraction column and develop a good material balance around this column, pilot plant testing using actual plant feed material is required.

The aqueous feed for this application was unique in that it was a distillate stream from a resin manufacturing process. The distillate stream was actually two phases; a phenol rich phase and an aqueous rich phase that contained the soluble amount of phenol which was analyzed at room temperature to be 7.4%. The phenolic phase by-passed the extraction column and was directed to the distillation column for phenol recovery in the bottom of the column. The aqueous phase was directed to the extraction column for counter current contact with MIBK. The target for the raffinate concentration was < 50 ppm phenol.

Prior to pilot plant testing the extraction process, LLE tests [7] with the actual feed were performed to generate LLE data and finalize the extractor configuration for pilot plant set-up. These tests demonstrated that 6 theoretical stages were required to achieve the <50 ppm phenol in the raffinate at S/F = 0.2 "as is" basis.

Pilot plant testing was performed utilizing a 3" diameter x 60-stage SCHEIBEL® Column as shown in Figure 4 below. Prior simulation for both this feed and previous phenol extraction systems, plus performance in actual commercial systems demonstrated the < 5ppm phenol could easily be achieved in the MIBK distillate for recycle to the extraction column. Thus, no phenol spiking of the MIBK solvent for the extraction tests was required. During testing, the design capacity for scale up was determined to be 600 GPH/ft^2. Based upon raffinate analyses, < 50 ppm raffinate phenol concentration could easily be achieved using 42 - 60 agitated stages and S/F = 0.2. The results indicated that lower S/F with 60 stages could also be used. In the end, the SCHEIBEL® extraction column was sized for 60-stages and S/F = 0.2 in order to be conservative. Like the other case studies discussed, process simulation software and distillation internals hydraulic rating software were used to design the phenol recovery and wastewater stripping columns. A complete modular system including all of the unit operations and other equipment, piping, instrumentation and controls was designed and fabricated designed and fabricated off site and delivered to the end user.

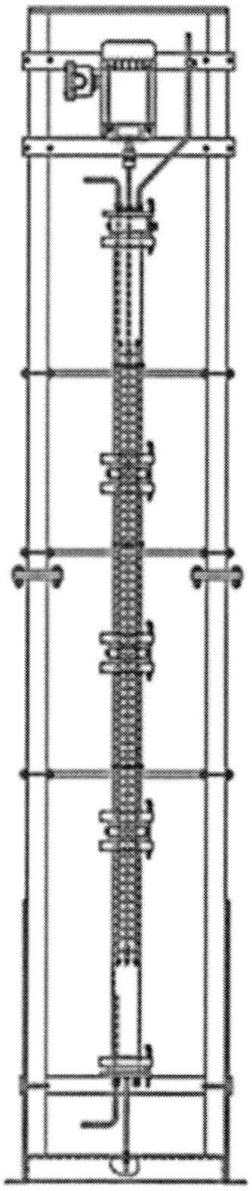

Figure 4 3" Diameter / 60 Stage Scheibel Column Used for Phenol Extraction Pilot Test

CONCLUSION

Designing a liquid-liquid extraction process with a solvent that boils lower than water is used to extract, and possibly recover, higher boiling compounds from aqueous streams requires optimized design of not only the extraction column, but also the interaction between the extraction column and the downstream distillation columns required for the entire system. The three case studies presented in the paper show that first it is important to select the correct extraction column technology and then to have some understanding as to the amount of solute that will be present in the recycled solvent. Pilot plant testing the extraction column is almost always required in order to accurately design the column itself and to determine the corresponding material balance, which in turn allows for downstream distillation design. When sufficient VLE data is available for the known components in the system and the design engineers have experience with process simulation and distillation internals design, then these downstream columns can be designed without the need for pilot plant testing. However, if VLE data is not known or if there are concerns with multicomponent interactions that may not be accurately simulated, or if there are other factors that inhibit confident distillation design, then pilot plant testing the downstream distillation columns can be very useful for their design and de-risking the entire process.

One note of caution, pilot plant testing in the manner discussed in the paper, often cannot account for possible build-up of low boiling contaminants in the solvent, especially when these components are in very low concentration in the aqueous feed. The only way to really determine this would be to set up a full pilot plant system with all the required columns operating in recycle mode and to operate this system for a sufficiently long period of time. Unfortunately, this is not really practical for most applications.

REFERENCES

1. Cusack, R., Karr, A., "A Fresh Look at Liquid-Liquid Extraction, ", Chemical Engineering, April 1991
2. Cusack, R., "Pilot Plants Confirm Process Validity", Chemical Engineering, June 1998
3. Glatz, D., Parker, W., "Recovery Of Valuable Products From Biomass Using A KARR® Column", International Solvent Extraction Conference (ISEC 2008), September 2014
4. Cusack, R., Fremeaux, P., A Fresh Look at Liquid-Liquid Extraction, The Range of Available Extractors is Wide", Chemical Engineering, March 1991
5. Glatz, D., Cross, B., Lightfoot, T., "Pilot Plant Testing of Liquid-Liquid Extraction", CEP Magazine, February 2018
6. Glatz, D., Parker, W., "Successful Design of Agitated Extraction Liquid-Liquid Extraction Columns Via Pilot Plant Testing", International Solvent Extraction Conference (ISEC 2014), September 2014
7. Glatz, D., Cross, B., Lightfoot, T., "Liquid-Liquid Extraction: Generating Equilibrium Data", Chemical Engineering, October 2018

592176 Invited Talk: Engineering a More Sustainable Polyurethanes Future

Abstract:

The Dow Chemical Company has championed sustainability for many decades. Furthermore, our mantra "Waste Reduction Always Pays" underscores a belief that sustainable solutions will ultimately be economically favorable. Chemical engineers play a key role in improving the sustainability of products manufactured in our company's plants. This presentation will describe four separate ways that engineers are increasing the sustainability of polyurethanes in an economically-advantaged manner. First, increasing the energy efficiency of the PMDI manufacturing process through mixing improvements will be discussed. Second, opportunities to alter reaction selectivity and reduce solid waste generation will be given. Third, the incorporation of CO_2 into polyol would capture and utilize a gas stream that could otherwise contribute to global warming. And lastly, the development of processes to recycle waste polyurethane foams directs us toward the establishment of a circular PU economy.

Bio for speaker: Dr. Paul Gillis, Fellow, Dow Chemical

After receiving a Ph.D. in Chemical Engineering from Brigham Young University in 1989, Paul joined the Fluid Mechanics discipline in Corporate R&D within the Dow Chemical Company. Paul pioneered the application of computational fluid dynamics (CFD) to design, debottleneck and develop new chemical processes. He is known for his fluid dynamics and mixing expertise and developing new tools for multi-phase flow with phase change and mixed Damkohler number reacting flow simulations. After becoming a Fellow in 2003, Paul has worked across a number of Dow business units including Polyurethanes, Epoxy, Oil & Gas, Emulsion Polymerization, and Electronic Materials. Outside of work, Paul is passionate about genealogy and camping and hiking in U.S. National Parks with his family.

589298 Development of a Single System for Converting Carbon Dioxide to Value Added Chemicals Using Circulating Fluidized Bed Technology

Abstract:

Kyle March, Eloy Flores, Chemical Engineering Dept., Division of Chemistry and Chemical Engineering, Southwest Research Institute, San Antonio, TX 78238.

Topic: Fluidized reactor technology /Carbon capture/CO2 utilization

Development of a single system for converting carbon dioxide to value added chemicals using circulating fluidized bed technology.

Current carbon dioxide emission reduction plans focus on mitigating emissions at the industrial process point source. Typically capture and separation process for carbon dioxide create a concentrated stream of carbon dioxide that may be used as a feedstock for chemicals or fuels. SwRI has been on the forefront of this effort for more than 10 years, and has continued to search and work with its clients for innovate ways of capturing, separating, and utilizing carbon dioxide. SwRI is working to develop catalyst and processes to convert carbon dioxide to value added products using temperature and pressure in a circulating fluidized bed reactor. By combining SwRI's knowledge and experience with circulating fluidized beds, catalyst development, and carbon dioxide conversion, a process for converting carbon dioxide to value added products is being researched. This presentation will discuss the ongoing research and development.

589680 Radio Frequency Driven Catalytic Reactors for Portable Green Chemistry

Abstract:

Heterogeneous catalytic reactors account for roughly 80% of all chemical manufacturing processes,[1] and are often driven by external heating of the catalyst via fuel-fired furnaces or utility steam. Traditional methods relying on combustion of fuel often lead to greenhouse emissions, limit distributed chemical production due to economy of scale, require additional infrastructure and large reactor size to accommodate combustion zones and insulation around the reactor. They also have thermal gradients in reactors which often compromises performance, utilization, lifetime, and selectivity of the catalyst.

The volumetric electric heating methods and modular reactors for power-to-chemicals route can pave a path towards distributed manufacturing and reduced greenhouse gas emissions when electricity is harnessed from renewable energy sources. Uniform volumetric heating of catalyst can also improve catalyst utilization, avoid homogeneous side reactions, and improve reactor portability.[2] Since the first published report of microwave organic synthesis, [3-4] microwave heating has been studied for various endothermic catalytic reactions.[5] However, high power microwaves (> 2.45 GHz frequency) are limited due to temperature hotspots, runaway reactions, penetration depth, reflection losses, and stringent safety exposure limits.

This study employs a multidisciplinary approach to make portable reactors by using novel materials like carbon nanotubes (CNTs) and silicon carbide (SiC) fibers as additives in catalyst. We utilize interaction of these materials with radio frequency fields (1 MHz-300 MHz) to selectively heat catalytic reactors and drive a reaction. A proof-of-concept is demonstrated for methanol steam reforming reaction using platinum as a catalyst. The RF heating response of CNT/platinum /alumina and SiC fiber/platinum were investigated for varying power using different kinds of noncontact RF applicators. The conversion of methanol for different reaction temperatures was compared to conventional ovens. This power to chemical method has application in modular reactors for on-site and on-demand production of chemicals using electric power.

(1) Fechete, I.; Wang, Y.; Védrine, J. C. The past, present and future of heterogeneous catalysis. *Catalysis Today* **2012,** *189* (1), 2-27, DOI: https://doi.org/10.1016/j.cattod.2012.04.003.

(2) Wismann, S. T.; Engbæk, J. S.; Vendelbo, S. B.; Bendixen, F. B.; Eriksen, W. L.; Aasberg-Petersen, K.; Frandsen, C.; Chorkendorff, I.; Mortensen, P. M. Electrified methane reforming: A compact approach to greener industrial hydrogen production. *Science* **2019,** *364* (6442), 756-759, DOI: 10.1126/science.aaw8775.

(3) Gedye, R.; Smith, F.; Westaway, K.; Ali, H.; Baldisera, L.; Laberge, L.; Rousell, J. The use of microwave ovens for rapid organic synthesis. *Tetrahedron letters* **1986,** *27* (3), 279-282.

(4) Tse, M. Y.; Depew, M. C.; Wan, J. K. S. Applications of high power micro wave catalysis in chemistry. *Research on Chemical Intermediates* **1990,** *13* (3), 221-236, DOI: 10.1163/156856790x00102.

(5) Ramirez, A.; Hueso, J. L.; Abian, M.; Alzueta, M. U.; Mallada, R.; Santamaria, J. Escaping undesired gas-phase chemistry: Microwave-driven selectivity enhancement in heterogeneous catalytic reactors. *Science Advances* **2019,** *5* (3), eaau9000, DOI: 10.1126/sciadv.aau9000.

Membrane Water Gas Shift Reactor Design: Applied to 10 MTPD Modular Ammonia Plant.

By Richard B. Strait

Abstract:

Extensive research is being performed to develop membrane reactors for hydrogen production. Much of the research is focused on the Steam Methane Reforming (SMR) reaction and to a lesser extent the Water Gas Shift (WGS) reaction. In parallel we have been designing a 10 metric ton per day (MTPD) modular ammonia plant which can be transported to sight as cargo and set up to operate using local feedstock to produce fertilizer for local consumption. If a low cost reliable membrane SMR or WGS or even Ammonia Converter were available they would allow for smaller equipment that did more making the ammonia process more cost effective and suitable for compact, small modular plants.

This paper focuses on one of these membrane reactors, the WGS. The basic design will be presented and the positive aspects of the design along with the challenges as they apply to a modular Ammonia Manufacturing Box (AMB) producing 10 MTPD next to the farmer's field.

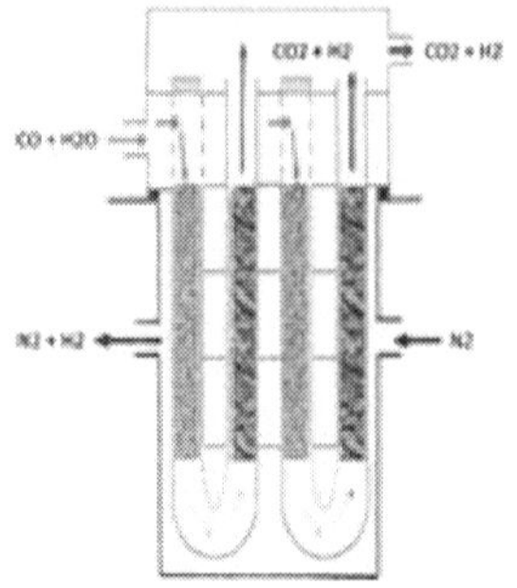

The mechanical design is discussed including:

- Membrane materials selection, cost and reliability
- Construction, maintenance, turnaround and emergency operation.

The process design is discussed including:

- WGS Reactor and catalyst performance
- Hydrogen separation performance
- Reactor temperature control for peak performance

RESEARCH TO APPLICATION

Research on how to produce hydrogen from natural gas, synthesis gas and carbon monoxide using membrane reactor systems is plentiful [5][8][10]. Most focuses on Steam Methane Reforming (SMR) [5] [8] and Water Gas Shift (WGS) [2] [3] [4] using palladium as the membrane material that allows hydrogen to pass through without other molecules. Research on small modular designs is also plentiful [6] [9]. The problem is, with all this research on how to generate hydrogen and ammonia using new leading edge technology, little has been implemented in an ammonia production facility. The ten metric ton per day (10 MTPD) modular design referred to as the Ammonia Manufacturing Box (AMB) includes many of these new and leading edge designs. The design focus is low capital cost, low energy consumption, safety and operability. The overall objective is to provide low cost fertilizer to the farmer's field produced from locally available raw materials.

AMMONIA 10 MTPD DESIGN

Ammonia is produced in hundreds of plants worldwide and is one of the largest commodity chemical globally at 160 million metric tons. The 1950's was the decade of the 100 – 200 mtpd reciprocal plant. Late in the 1960's and 1970's was the period of the MW Kellogg large centrifugal ammonia plant (600 – 1000 mtpd). Also in the 1970's CF Braun invented the Purifier plant with low energy performance at 1000 mtpd. MW Kellogg was the dominate player in the 1970's and into the 1980's with replicated plants in the 1000 or 1500 mtpd size [2]. Starting the 1980's energy consumption was the primary focus when natural gas prices were increasing all the competitors adjusted their designs to reach a 7.0 Gcal/MT energy point where it stands today. CF Braun had the clear advantage by having a low energy design already in service. With all the ammonia licensors having the same energy consumption level the next attack was on capital cost. The economy of scale on a per ton basis can result in a 30% cost savings if properly applied up to the limit of parallel trains. This lead to ever larger Mega Ammonia plants in the 2000's and 2010's. Today the low end of the capacity range is 2000 mtpd and the largest 3300 mtpd with the cost per daily metric ton of nameplate capacity about $250,000 for the ISBL. The supporting offsites might cost another $125,000 for Brown field while Greenfield $250,000 or even more. Large plants cost more than $US 1 billion but on a per ton basis are considered the most cost effective. Today however the micro plant is economical and cost effective if the latest technology is applied keeping the CAPEX at $500,000 per daily ton for the complete plant and the energy consumption at 7.0 Gcal per ton.

Since 2000 most of the Rice University Chemical Engineering Senior Design Projects (Capstone) have been to design a synthesis gas chemicals plant with various feedstocks, at varying sizes, in remote locations. Sustainably, Green, Process Intensification, Modular, Mobile, Micro Processing Technology, Local designs for Local needs, Renewables, Latest Technology and Creativity were all themes used in making the Capstone project problem statement. Many of those projects sifted through the current research and found the cost effective new technologies around hydrogen production, ammonia production and modular designs.

SMALL MODULAR ADVANTAGE:

The overall objective is reduce the cost of fertilizer to the farmer in remote locations. Mega plants are the low cost producers in most cases but only for those located within a reasonable shipping distance to the end user (farmer). With the focus on the farmer in a remote location the Mega Ammonia plant is just not a good fit. In many locations the transportation of ammonia to site is difficult and costly plus in many poorer countries there are other related costs added to the price before it gets to the farmer's field. For example a Tanzanian farmer might be able to buy ammonia for $500/MT at the plant gate in Qatar but that price then has to be increased to cover ocean and land transportation ($50 est.) to the farmers field and other costs such as import duties, tariffs, currency adjustments, wholesalers and retailers mark ups

($150 est.). The Qatar plant gate price of $500/MT grows to $700/MT or more at the farmer's field. Several countries subside the fertilizer price to the farmer to offset this higher cost and at the same time subsidize food prices. In developing countries food supply and price are important issues along with balance of payments.

The 10 MTPD modular ammonia plant that can be installed for $5 million ($500,000/DMT) operate consuming energy of 7.0 Gcal/MT while using locally available raw materials, natural gas, air and water.

EQUILIBRIUM REACTORS: THE MEMBRANE ADVANTAGE

Ammonia plants have 7 reactor systems: Sulfur Removal, Primary Reformer, Secondary Reformer, High Temperature Shift, Low Temperature Shift, Methanator and Ammonia Converter. Of those 7 all but 2 are good candidates for membrane reactors because they are equilibrium reactions.

SMR is a good choice as it is central to producing hydrogen from reforming natural gas and/or CO_2.

$$CH_4 + H_2O \leftrightarrow CO + 3H_2 \qquad (+ 206\ kJ/mole)$$

The SMR as a membrane reactor is challenging because the reforming reaction occurs at high temperature (900 C) and a high heat flux is required to feed the highly endothermic reaction (+206 kJ/mole). It has not been ruled out but other options are being considered.

WGS is a good choice as it completes conversion of feedstock to the maximum amount of hydrogen in a range of lower temperatures (450 C to 200 C).

$$CO + H_2O \leftrightarrow CO_2 + H_2 \qquad (- 41\ kJ/mole)$$

For the 10 t/d plant this looks to be the best application of a membrane reactor.

The ammonia converter is a good choice as an equilibrium exothermic reactor but at the converter pressure (125 bar) cooling and condensing of the product stream to remove ammonia is preferred as less costly.

$$N_2 + 3H_2 \leftrightarrow 2NH_3 \qquad (- 92.44\ kJ/mole)$$

The ammonia converter will be a new micro design that keeps the operation at or close to the equilibrium curve though out its length.

WGS MEMBRANE REACTOR:

Key mechanical features:

a) Removable Bundle
b) Double Tubesheet
c) Partitioned Shellside
d) Membrane Tubes
e) No Internal Flanges
f) Reaction, Separation and Heat Transfer all in one unit replacing several items.

Key process parameters:

a) Pressures in line with existing ammonia plants but with higher pressures where it helps the micro size plant overall efficiency and capital cost.
b) Temperatures closely tracking the equilibrium curve and close approaches in the heat transfer equipment.
c) Flows within the tubes and outside that match the best performance for the reactor conversion, membrane separation and heat transfer.
d) Compositions that complement the rest of the unit. Hydrogen directly to the process nitrogen stream. More complete conversion of CO.

In an ammonia plant there are two shift reactors, high temperature shift (HTS) with a high reaction rate and low temperature shift (LTS) with a close approach to equilibrium at a low temperature consuming most of the CO and producing maximum hydrogen.

Some ammonia plant designers have offered a medium temperature shift (MTS) as a means of lowering the CAPEX of the plant by eliminating equipment items. This has not caught on because the operations within the two shifts HTS and LTS are so different with respect to catalyst performance, type and temperature of operation. The micro plant at 10 mtpd will have a precision temperature shift (PTS) where the temperature is closely controlled along the length of the tube containing catalyst. Plus the catalyst at each temperature step will be suited specifically for that temperature and concentration of reactants.

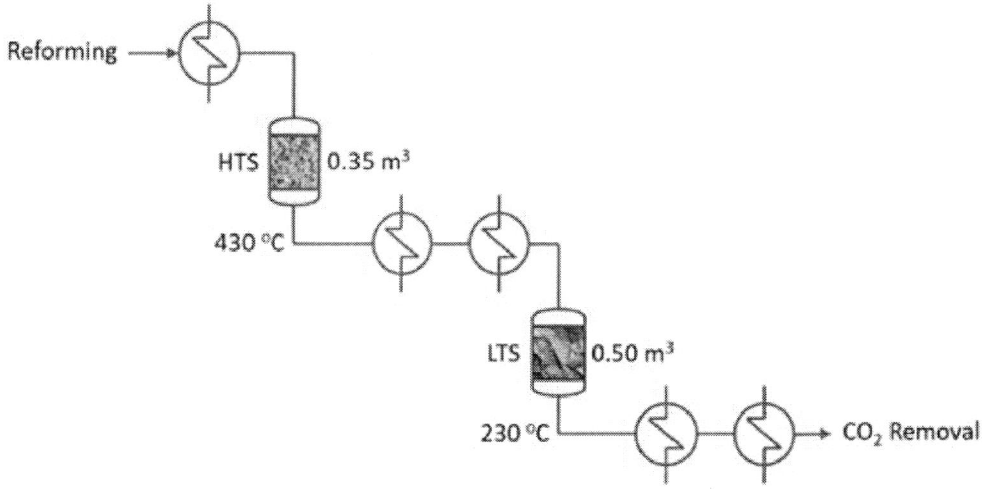

Current PFD at 10 MTPD

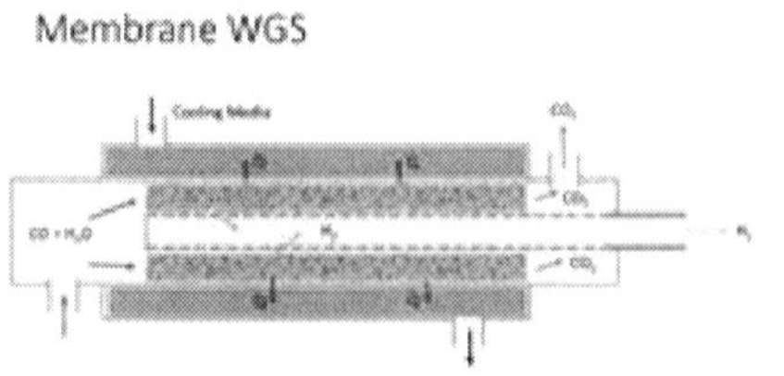

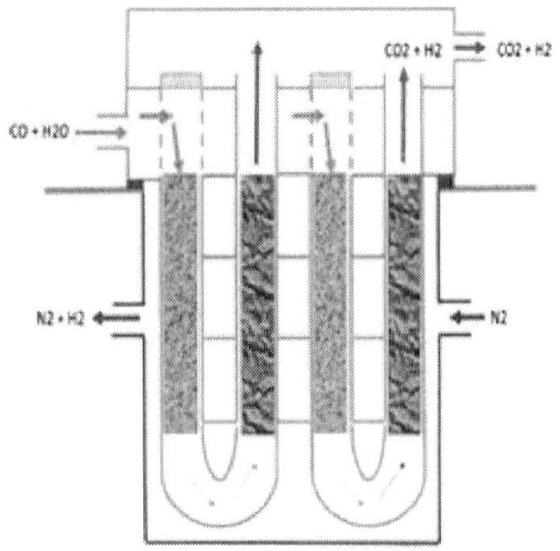

Single Tube Membrane WGS

Multi Tube Membrane WGS

Our starting point was two smaller reactors with the same performance to the normal HTS and LTS, but on a micro scale and using modern technology these adiabatic reactors can be combined into one precise temperature controlled membrane reactor containing proportionally less catalyst and a precision temperature match for each catalyst along the reactor tube. Ultimately the new design saves catalyst, equipment and energy for the process. Hot syngas will enter the tube where the reaction will occur rapidly. The heat generated from the reaction passes from the tube through the wall to the nitrogen sweep gas on the shell side. The tube wall is also the membrane allowing hydrogen to preferentially pass through it into the speed nitrogen on the shell side. The shell side of the MWGS contains nitrogen which will be used in the ammonia reaction later. This nitrogen is not contaminated by the hydrogen but actually enriched as it builds up to the required 3:1 ratio required for ammonia production.

MEMBRANE SELECTION:

Palladium is widely studied as the membrane material because it provides the best separation (100% pure H2) but it is very expensive at 1800 – 2800 US$ per ounce. Palladium is the most effective membrane material because it uses proton transfer to allow the hydrogen molecule to disassemble on the tube side and reassemble on shell side with almost perfect separation of the hydrogen at 100% purity. But Pd adds $75,000 - $150,000 to the cost of the small WGS reactor in our small plant. Palladium also performs well at 450 C but at 400 C the permeation rate is cut in half and at 350 C the CO starts to block the hydrogen from permeating through the tube [2]. Nickel has been tried as a lower cost alternative but it is only 8% effective as Pd in separating the hydrogen [3]. Ceramics and zeolites are promising [11]. At this point zeolites are the most promising as being most overall cost effective. The membrane and precise temperature control will allow us to reduce the size of the WGS reactor by 25 – 50%.

MECHANICAL DESIGN:

The WGS takes place in a porous catalyst filled tube with nitrogen flowing on the outside of the tube which sweeps away the permeate hydrogen while cooling the reaction. This u-tube heat exchanger reactor has a removable bundle facilitating simple field change out of the catalyst charge. The u-tubes allow the tubes to grow and shrink without binding and a two tubesheet design allows all the connections to be located at the top of the reactor/separator/exchanger.

PROCESS DESIGN:

The process design takes the research papers into the process plant instead of shrinking an existing design 200 times. Process Intensification is central to the design leading to a modular design that can be shop fabricated and transported to site as a single unit. Also with a one size fits all design it can be replicated a 1,000 or 10,000 times to lower the capital cost per ton of production.

Capital Cost (CAPEX) is reduced at every turn. The catalyst is more effective in our WGS membrane reactor/separator/exchanger. Equipment is combined and now new improved catalyst formulations are possible. But it goes well beyond the WGS into every process portion of the plant including integrated utilizes.

Operational Costs (OPEX) is also as low as 7.0 Gcal/MT or even lower depending upon how far we go with equilibrium reactor designs and other equipment within the plant.

SAFETY AND OPERABILITY:

The operation of the plant will be by farm center operators with extreme instrumentation and monitoring by computers and a minimum of operator attention required. The operator is there to keep things running though maintenance activities and reporting activities and to shut down the plant if required because the instrumentation did not. The operator is not there to make operational moves but rather to shut down and secure the plant if necessary. Safety is built in with a goal of minimum operator interface, maximum monitoring and automatic adjustment and shutdown if necessary to a safe condition.

SUMMARY:

- No significant change to the ammonia process design since the 70s (centrifugal plant) and 80s (low energy plant). Since the 60s and though to today all licensors have been making ever larger plants to take advantage of the economy of scale to build lower cost per ton plants but that has its limits at 2000 – 3000 MTPD.
- The 10 MTPD plant is modular and takes process intensification to its practical limit, using membrane reactors with integral heat and mass transfer.
- Membrane material will be ceramics/zeolites. Palladium would add $75,000 to $150,000 to the cost of the membrane WGS and is considered too expensive.
- The WGS is described as a system ready to be integrated into the plant and beta tested.
- The operation of the plant will be by farm center operators. Experts are brought in as needed.
- Safety will be enhanced because of addition instrument and computer operational control.

REFERENCES

[1] Appl, Max, *Ammonia: principles and industrial practice.* 1999. Print.

[2] Alexander S. Augustine, Yi Hua Ma, Nikolaos K. Kzantzis, "High temperature palladium membrane reactor for high temperature water – gas shift reaction", *Intl. Journal of Hydrogen Energy 36* **2011**, 5350 – 5360 [X002]

[3] K. Babita, S. Sridhar, K.V. Raghavan. "Membrane Reactors for fuel cell quality hydrogen through WGSR – Review of their status, challenges and opportunities." *Intl. Journal of Hydrogen Energy 36* **2011**, 6671-6688

[4] M.D. Dolan, R. Donelson, N.C. Dave, "Performance and economics of a Pd-based planar WGS membrane reactor for coal gasification." *Intl. Journal of Hydrogen Energy 35* **2010**, 10994 - 11003 [X003]

[5] Abu El Hawa, H. W., Lundin, S., et al. "Steam methane reforming in a Pd-Au membrane reactor: Long-term assessment" *Intl Journal of hydrogen energy* **2016** *41*, 10193 – 10201 [C003]

[6] Jensen, Klavs F. "Silicon-Based Microchannel Systems: Characteristics and Applications." *MRS Bulletin* **Feb 2006,** *Vol. 31,* 101-107.

[7] Multhaup, Robert H. and Eschenbrenner, Gunther P., *Technologies Harvest, Feeding a Growing World Population.* 1996. Print.

[8] Simakov, David S. A. and Sheintuch, Mosche. "Experimental Optimization of an Autonomous Scaled-Down Methane Membrane Reformer for Hydrogen Generation." *Ind. Eng. Chem. Res.* **2010,** *49,* 1123-1129.

[9] Vaccaro, S., Malangone, L., Ciambelli, P. "Micro Scale Catalytic Reactor for Syngas Production." *Ind. Eng. Chem. Res.* **2010,** *49,* 10924-10933.

[10] Yan, Y., Cui, Y., et al "Experimental investigation of methane auto-thermal reforming in hydrogen permeable membrane reactor for pure hydrogen production." *Intl Journal of hydrogen energy* **2016** *41*, 13069 – 13076 [C002]

[11] Yonamine, W., Thangavel, S., et.al. "Performance Analysis of a water gas shift membrane reactor for integrated coal gasification combined cycle plant." *Energy Conservation and Management 174* **2018,** 552-564 [C001]

R. B. Strait
Adjunct Professor Rice University = RBS1@rice.edu
Ammonia Consultant = Richard.Strait.Syngas@gamil.com
February 26, 2020

593632 Sponsored Technology Workshop - Extract More Value From Your Existing Simulation Software Today - Chemstations

Abstract:

Are you using your simulation software as much as you could be? Come learn the collected wisdom that can increase the return on your process simulation investment. We see three significant areas of benefit: (1) insure that you and your colleagues are trained to use the fundamental features of a simulator, (2) make certain that everyone who can benefit from simulation has access, and (3) raise awareness of advanced features in modern simulators. Companies that encourage simulation for problems of all sizes and allow access by employees in all process-related groups can achieve greater bottom-line benefits with the investment they are already making.

In this interactive workshop, we will cover tips and tricks to help get the most out of any simulator for typical applications and an overview of some advanced features that can bring significant benefits to your bottom line. After a short presentation, we will split into virtual breakout sessions covering demonstrations of specific advanced topics. You will find this workshop material beneficial no matter which vendor's process simulator you have in your organization.

Don't miss this unique opportunity to ask questions, meet industry peers, and add to your process simulation knowledge.

590910 Industry-Academic Collaboration for the Transition to a Low Carbon World

Abstract:

New technologies solutions and novel ways to develop them are needed to solve the energy challenges that the world is facing. Engineers and scientists of our time will play a critical role to provide energy, fuels, and chemicals with low carbon footprint to a growing world population. The magnitude of the task is such that no one can answer these technical challenges alone and a sustained collaboration is needed between industry, academic institutions, and national laboratories. A key for success is to understand how to leverage the diverse perspectives that industry and academia bring, and how to manage different drivers and goals of each parties. In this talk we will reflect on the role of open innovation in navigating the energy transition, and how collaboration is critical in selecting and tackling problems worth solving.

589933 Using Open Innovation to Design Chemical Products

Abstract:

In this presentation we will share our experiences as a solution provider for several open innovation projects. The goals of these projects were to develop new chemical products ranging from novel automotive windshield washer fluids to advanced nail polish formulations. Some of the experiences we will discuss include: how we use the NineSigma Open Innovation platform as our "matchmaker" for connecting our small business, having an expertise in physical property modeling, with large companies seeking solutions; how Q&A opportunities before proposal submissions can be extremely helpful; how having a champion in the company seeking a solution is very important. Finally, we will show why we believe the Open Innovation model has provided us with opportunities to collaborate with businesses whom we might never have thought could use our expertise.

590133 Ncerc at Siue: A Hub for Open Innovation in Biofuels and Bioproducts

Abstract:

The National Corn-to-Ethanol Research Center at Southern Illinois University Edwardsville (NCERC at SIUE) is a nationally-recognized research center dedicated to the development of biorenewables: biofuels, biochemicals, biomaterials and bioproducts. Since 2003, NCERC has grown with the biorenewables industry through a complete portfolio of services in the bioeconomy. The NCERC offers a modular, plug-and-play facility design for optimum process and experimental flexibility, convenient, in-house scale up capabilities from lab to intermediate to pilot scale, and employs leading scientists, engineers and operators to take your project to the next level. The NCERC maintains strong client confidentiality and intellectual property protection, allowing flexibility for clients, along with a culture of leadership and customer service.

The NCERC's resources are cutting edge. We realize that in a fast-paced industry, it's not only what you know, but who you know. The NCERC has deep relationships with leaders across several key sectors: industry, academia, government, and trade associations. Our network of information and resources are there to support clients and their project.

With no core funding for instrument maintenance or workforce support, the NCERC is constantly faced with challenges. Communication with clients is critical, especially for long-term projects without a client presence on site.

When it comes to research, we are on this journey together. What the client envisions, NCERC achieves. As a platform for open innovation by clients, NCERC delivers superior results through adaptable, innovative thinking, supported by our plug-and-play facilities and unparalleled expertise in engineering design, method development, process control and automation, strong client confidentiality and intellectual property protection, and commercialization services: project design, proof of concept, validation and scale up.

589281 A Porous Copper Metal Organic Framework with Highly Nucleophilic Amine Links for CO_2 Gas Separation from Biogas

Abstract:

A porous copper metal organic framework with highly nucleophilic amine links for CO_2 gas separation from biogas

Kritika Narang[1], Andreas Kaiser [2], *Farid Akhtar* [1*]

[1] *Luleå University of Technology, Luleå 97187, Sweden*

[2] *Technical University of Denmark*

*: *corresponding author, email: farid.akhtar@ltu.se*

The rising interest in using biogas as an energy fuel has lowered the consumption of fossil fuel and inclined the environment toward sustainability. Despite having low CO_2 emissions during the combustion of biogas, it still involves a significant challenge to remove the existing impurities in the biogas using the most efficient technology before becoming a substantial fuel for the industries. Biogas consist of ~ 40 mol% of CO_2 depending upon the substrate and conditions of the digester used during production [1]. This CO_2 is the major undesirable component for the energy recovery of the biogas as it lowers the calorific value of the fuel. Adsorption driven materials have set a benchmark in CO_2 separation in the biogas upgrading process [2,3]. Wide variety of adsorbents has been studied for their separation performances and energy consumption to procure efficient, energy-saving and environmentally benign separation processes [4]. MOFs (Metal-organic frameworks) are porous materials, which has gained attention in the past two decades and has widespread potential applications in gas separation and purification technology. MOFs offer unique properties such as high surface area, tunable pore size and easy tailoring of chemical functionalities.

In this research, we report a copper MOF synthesized using a facile solvothermal synthesis. The Cu-MOF crystals were of cuboid-shaped of 300-500 nm in size and demonstrated a BET surface area of 756 m^2/g. The Cu-MOF showed carbon dioxide uptake capacity of 2.1 mmol/g at 273K and 100 kPa and 1.8 mmol/g at 293 K and 100 kPa. The gravimetric high-pressure adsorption was carried out for both CO_2 and CH_4 gases at 10 bar. The Cu-MOF showed 3.7 mmol/g uptake of CO_2 and 2.5 mmol/g for CH_4 at 10 bar and 293 K. The isosteric heat of adsorption (Q_{st}) using Van't Hoff equation was 18 kJ/mole for CO_2 adsorption, which describes the low energy penalty for the regeneration cycles. To investigate the CO_2 separation performance from CH_4, the Cu-MOF powders were structured into mechanically strong hierarchically porous pellets and breakthrough adsorption and desorption cycles were measured. The breakthrough data was used to obtain mass transfer coefficient and diffusivities value.

References

[1] I Angelidaki, L Treu, P Tsapekos, G Luo, S Campanaro, H Wenzel, et al., Biogas upgrading and utilization: Current status and perspectives, Biotechnology Advances. 36 (2018) 452-466.

[2] S Chaemchuen, NA Kabir, K Zhou, F Verpoort. Metal-organic frameworks for upgrading biogas via CO2 adsorption to biogas green energy, Chem.Soc.Rev. 42 (2013) 9304-9332.

[3] B Yuan, X Wu, Y Chen, J Huang, H Luo, S Deng. Adsorption of CO2, CH4, and N2 on Ordered mesoporous carbon: Approach for greenhouse gases capture and biogas upgrading, Environ.Sci.Technol. 47 (2013) 5474-5480.

[4] K Zhou, S Chaemchuen, F Verpoort. Alternative materials in technologies for Biogas upgrading via CO 2 capture, Renewable Sustainable Energy Rev. 79 (2017) 1414-1441.

590381 Influence of Particles on Amine Losses during CO_2 Capture: A Process Simulation Coupled Aerosol Dynamics Model

Abstract:

Efforts to reduce the concentration of atmospheric CO_2 have been initiated due to escalating concerns of global warming and climate change. Amine-based post-combustion CO_2 capture offers an economically viable retrofit for power-plants to help combat these emissions. However, solvent losses have been a major challenge to commercializing this technology. Recent experiments[1] show that vapor-based and aerosol-based solvent losses are the key mechanisms of amine solvent loss during CO_2 capture. While vapor-based amine losses can be abated by employing techniques like water-washing, there has been little work to curtail aerosol-based solvent losses. This is primarily because the underlying mechanisms governing these losses are still not well understood. Although there are a few numerical models proposed for this system, there are no models that consider aerosol phenomena such as coagulation or particle loss. Moreover, these models have not been fully validated through experiments.

In this work, a process simulation model for multi-component mass and heat transfer, hydrodynamics, thermodynamics, and electrolytic-solution chemistry was coupled with a multi-component aerosol dynamics model accounting for condensation, coagulation, and particle loss. The model's outputs were validated using the experimental data of Khakharia et al., 2013[2]. Based on the results, it was shown that neglecting particle losses inside the absorber resulted in over-prediction of amine-based solvent losses, while including coagulation resulted in reduced (~10%) emissions. Also, amine emissions increased with higher particle number concentration and lower initial mean particle size. Furthermore, it was shown that amine emissions decreased with lower solvent concentrations and temperatures, but CO_2 capture efficiency dropped for decreasing solvent concentration while remaining unchanged for lower temperatures. With this model, optimum operating conditions are suggested to reduce solvent losses during amine-based CO_2 capture.

References

1. International Journal of Greenhouse Gas Control, vol. 85, pp. 86-99, 2019.
2. International Journal of Greenhouse Gas Control, vol. 19, pp. 138-144, 2013.

590276 Integrated Mid-Scale Natural Gas Liquefaction Solution- Factors That Can Lead to a Viable Mid-Scale Project

Abstract:

Mid-scale LNG is generally defined as a single or multiple LNG train(s) with capacity of 1.0 to 3.0 MTPA. The main factors supporting this development is a shorter schedule and certainty of cost by using standardized designs. The mid-scale concept is being applied to both the revamp and green field projects.

The Mid-Scale LNG projects are conceptualized to realize benefit of handling manageable pieces thus reducing the risk of cost and schedule overruns. The mid-scale solution not only brings more vendors and contractors available in this space, but it also allows for more design alternates and thus more flexibility during the project execution thereby reducing the overall risk significantly. In some cases, depending on the guiding factors, multiple mid-scale trains can be a potential viable alternate for a LNG project.

The paper discusses the integrated mid-scale LNG solution in detail and its advantages in certain conditions. The integrated solution entails an off the shelf standardized pre-engineered design using the APCI SMR / DMR technology for Liquefaction and Technip Energies design for Pre-Treatment section. The paper discusses few case studies along with risk based economic analysis and illustrates the factors, if realized, that can lead to a viable Mid-scale LNG development in certain cases.

590092 New Modeling Approach of Mixing Quality and Axial Hold-up Distribution in Agitated Liquid-Liquid Flow Mixers for Enabling Process Intensification with Validation Using Advanced Techniques

Abstract:

Liquid-liquid mixing, extraction, reaction, separation, and beneficiation have found a wide range of industrial applications. New theoretical based models have been developed and validated to enable process intensification by properly predicting the mixing quality and the local dispersed phase hold-up distributions in the axial direction of a continuous liquid-liquid flow mixer. The models relate the mixing index and axial hold-up distributions to a Peclet Number, the average dispersed phase hold-up, the input flow rates and the physical properties of the dispersion. The developed models offer a new understanding of the nature of the mixing process in continuous mixers with two liquid-liquid immiscible phases in flow. Experimental work was also performed to investigate the mixing index and axial dispersed phase hold-up distribution using advanced Laser Transmission Technique (LTT) in a mixer setup. The experimental results were used in validating the models. The experimental and the predicted data agreed well for a wide range of impeller speeds/flow ratios. Furthermore, advanced Gamma-ray based visualization measurement techniques can be integrated to validate the model in an intensifying way.

589564 Commercial Experiences and Further Development of WINTRAY® - High-Throughput and High-Efficiency Liquid-Liquid Extraction Column -

Abstract:

Liquid-liquid extraction is a separation process for recovering the target components in the feed by contact with the solvent. Liquid-liquid extraction is a versatile technology with a wide range of commercial applications. It is used in the processing of refinery, petrochemical, coal and biomass-derived chemicals, complex organic compound production such as pharmaceuticals and nonferrous metals production. In addition, liquid-liquid extraction also is becoming important in industrial wastewater treatment due to the increase of environmental regulations in the world.

In designing the process which uses liquid-liquid extraction, the type of extractor needs to be selected from several kinds of equipment. The selection is very important in capital cost because the size of the extractor and required plot area varies by the type of extractor. The type of extractor also effects the required solvent circulation ratio and the energy consumption for the regeneration in the distillation section. In addition, the maintainability varies with type of extractor. These are directly linked to the operation cost of the facility.

The WINTRAY® which is liquid-liquid extraction column, has been developed to obtain a high-throughput and high-extraction efficiency. The throughput is 2-5 times higher than conventional equipment depending on the application. Conventional liquid-liquid extraction equipment is described below to compare with WINTRAY® .

Mixer-settler can achieve high stage efficiencies because of its long residence time. And it can handle high-viscosity liquids. The operation range is relatively wide because the rotation speed of the agitator is adjustable in the field. It is used in petrochemical, nuclear and nonferrous metal fields. But it requires huge plot area especially when a lot of stages are required. In addition, the equipment and maintenance cost are relatively high because it comes with the several number of agitators.

Rotating disc contactor (RDC) is also widely used in petrochemical, chemical, pharmaceutical and nonferrous metal fields. RDC comes with several theoretical stages in a single column with single agitator, therefore the equipment cost, and operating cost are moderate. And the rotating speed of the agitator can be adjustable in the field like as mixer-settler. On the other hand, because its throughput is relatively lower compared with other extraction column, the diameter of the column tends to be huge. And due to the difficulty in preventing the back mixing in the column, scale up and diameter of the extraction column are limited. In addition, the equipment cost is higher than static extraction column because it has agitator.

Sieve Tray which comes with multi perforated-plates, is relatively effective respect to both capacity and extraction efficiency of the extraction column. It also provides the several theoretical stages in single column like as RDC. It is static equipment, therefore, the equipment cost and maintenance cost are relatively lower than those of RDC. The scale up is relatively easy, and it is used in refinery, petrochemical and chemical industrial fields. However, because of the demand for increased petrochemical production and for cost reduction of the extraction process, further improved throughput and extraction efficiency for the extraction column has been desired.

The WINTRAY® which is liquid-liquid extraction column, can obtain a high-throughput and high-extraction efficiency. It combines two kinds of trays, composed of plates and weirs with unique shape openings, arranged alternately. The heavy liquid is dispersed into the light liquid in the column by the openings on weirs. After dispersed phase liquid accumulates on trays, it flows through the openings in each weir and then changes to liquid jets and makes several dispersed drops. The dispersed drops are larger than those formed in conventional columns, and shape is also different. The sequence of drop formation, movement and coalescence is achieved as the dispersed liquid moves from one tray to the next. In addition, higher frequency of the series involving dispersion, movement and coalescence can be attained with closer tray spacing. Therefore, the use of the WINTRAY® extractor results in both higher throughput and higher efficiency at the same time.

WINTRAY® was commercialized to obtain high capacity, high efficiency and fouling resistance in the liquid-liquid extraction. It has been already used in 19 commercial plants worldwide for various kinds of applications and conditions. The interfacial tension and density difference between heavy phase and light phase in commercial experiences cover the wide range of 1-35 mN/m and 20-600kg/m3, respectively. It means that WINTRAY® can be applied to various kinds of liquid-liquid extraction in industrial use. In addition, the problem of fouling has not been reported.

If WINTRAY® is applied for new construction of extraction column, the size of the column becomes compact because of its high throughput and high efficiency. High throughput makes the column diameter smaller, and high efficiency reduces number of trays and height of the column. In addition, as compared to extractors coming with driving device, the simple structure which consist of trays and weirs with openings, leads to low capital cost and maintenance cost.

When revamping the existing extraction columns, throughput can be increased only by replacing the internals with WINTRAY®. And also, the energy consumption in the solvent regeneration in distillation section can be saved by reducing the solvent circulation rate of the extraction column because of the high tray efficiency.

As an example of the performance of WINTRAY, the application to aromatic extraction column is described. In a commercial project, the perforated tray in the existing column was removed and WINTRAY® was installed in the aromatic extraction column uses sulfolane as a solvent. The number of trays and tray spacing were not changed after the replacement. After the modification, the amount of solvent is decreased 10% after installing WINTRAY® , and the BTX loss is reduced by 42% because of higher efficiency than the perforated tray, even though the capacity was increased 17%. In this project, the incensement of the capacity was limited because of the capacity of other facilities around the extraction column, and the actual capacity of WINTRAY® is higher than that. In the pilot test results, the throughput of WINTRAY® is approximately 40% higher than that of sieve tray.

In another commercial project, WINTRAY® was newly constructed in parallel with the existing RDC which is used for aromatic extraction. To reduce the aromatic concentration in the raffinate, the raffinate which contains less aromatic obtained from WINTRAY® , was mixed to the raffinate from the RDC and supplied to the downstream process. When using only RDC, the lower limit of the aromatic concentration in the raffinate is approximately 6%. On the other hand, it was reduced to 3% by parallel operation with WINTRAY® because of its high efficiency. In addition, even though the capacity per column cross-section area of WINTRAY® is double higher than that of the RDC, and the amount of solvent is decreased 20%, BTX loss is saved 70%. The height of the extraction part of the

WINTRAY® is 15% lower than RDC. We herein report the characteristics and review example cases of the usage and the advantages of WINTRAY® in detail.

We also introduce further development to improve the throughput of the extraction column. The newly developed tray is composed of novelty shape plates and weirs with openings.

In addition to the introduction of performance improvement, the prediction method of the flooding velocity is also developed. The flooding velocity is a key factor for deciding the diameter of the extraction column in new construction project, and deciding the capacity in the internal replacement project. The liquid-liquid extraction is a versatile technology with wide range of applications. The feed and operation conditions vary widely, especially in refinery industry. Therefore, the accurate prediction of the flooding velocity is very important when the technology is applied to new application and new conditions. We introduce the comparisons between the experimental results and the results which are estimated by our method.

Realizing the promised benefits of process intensification and integration through proper controller design

Michael A. Taube, S&D CONSULTING, INC., Houston, Texas, USA
Isuru A. Udugama Ph.D., Technical University of Denmark (DTU), Lyngby, Denmark
Brent Young, Ph.D., University of Auckland, Auckland, New Zealand
Christoph Bayer Ph.D., Technische Hochschule Nürnberg (TH-N), Nürnberg, Germany

Introduction

Process integration (mass and energy integration) is predominantly established as means for productivity improvement and cost minimization. The domain expertise is subject to development with active research being carried out into the topic. Currently the concepts of process integration are applied in the design of traditional petrochemical processes [1–3], while research has been carried out to also include process intensification objectives (usually reducing capital and operating costs by combining unit operations, functions or phenomena) to areas such as pharmaceutical manufacturing [4] as well as biotechnology based production processes [5]. Because of this diverse domain of application, process integration and intensification (PII) has been applied in numerous large-scale systems such as industrial (petro)chemical distillation units to less abundant applications such as micro-reactors.

Lutze et al. [6] defined process intensification as "a process development/design option which focuses on improvements of a whole process by adding/enhancing of phenomena through integration of unit operations, integration of functions, and integration of phenomena and/or targeted enhancement of a phenomenon within an operation". There is an increasing interest in application of intensified and multi-functional processes in chemical industry [7]. Several applications of process intensification principles are realized so far on an industrial scale including reactive distillation, micro-reactors and rotating packed bed systems, to name a few.

From a process design point of view, PII enables multiple operations or phenomena to be combined in a single unit operation while providing opportunities for identifying novel more sustainable process alternatives. From an economic point of view, intensified unit operations potentially offer multiple benefits including, but not limited to:

- Reduced capital cost by reducing the equipment required to carry out product synthesis and refinement [8]
- Reduced energy and operations cost due to better utilization of heat and materials flows [9], e.g. in baffled reactors
- Increased production due to the continuous in-situ removal of inert or inhibitory material [5], in case of combining reaction and separation in one-unit operation, e.g. reactive distillation.

Due to the potential benefits of PII, it has been adapted into a variety of commercial applications; examples include side draw distillation columns (SDC), divided-wall column (DWC), reactive distillation columns (RDC) and the in-situ removal of inhibitory molecules in bioreactors. However, despite the potential benefits, design and control of these intensified and integrated processes can be complicated due to loss in degrees of freedom resulting from combining operations, tasks or phenomena. It has been the author's experience that intensified processes in general are poorly controlled, due to the lack of process understanding that is caused by complexities brought on by PII. This paper describes a handful of case studies where additional complexities in terms of process control are outlined. Each case study also explains how the complexities brought on by PII were successfully tackled. The final section deals with lessons learnt that describe in general terms challenges faced and the ways to tackle potential pitfalls of control with respect to PII.

Success Stories

Side Draw Distillation Tower

The first example is a distillation column that performs the final clean-up of a C4 alcohol product at a specialty chemicals plant, removing both a light and heavy impurity from the final product; the purified C4 alcohol is taken as a side stream from the tower. This plant was built in early 1970's. Over time, a few of the original instruments – namely the Bottoms Accumulator Level and Bottoms Flow meter – were either abandoned or removed because of unreliability (due to fouling service). Control structure issues also affected the operations of the tower.

The original controls design on the overheads system had the Overhead Temperature manipulating the very large Reflux Flow, and the Accumulator Level manipulating the very small Distillate Flow (as usually found with "classic refinery control" designs); refer to Figure 1. This configuration never worked as the Reflux Flow was 2.5-3.5 times the Feed Flow, compared to the Distillate Flow which was only 2.5% of the Feed. Thus, the Accumulator Level, Overheads Temperature and Distillate flow constantly oscillated (but not due to poor controller tuning)

Figure 1: Side Draw Tower Original Control Structure

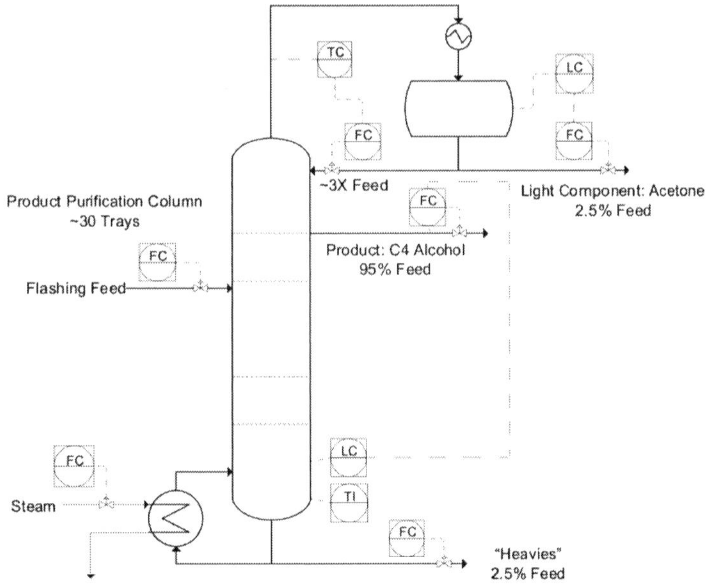

When starting the investigation, the tower was operated in a "semi-batch" style: The Feed was held constant and the overheads and Side Draw controls were operated in fully automatic/cascade control, but with the very poor behavior noted above. The Bottoms, however, was operated manually, as follows: The Bottoms Flow was blocked-in until the Sump Temperature High Alarm was reached (around 300°F); then the Bottoms Flow would be unblocked. The Sump Temperature would precipitously, and with very little delay, fall, at which point the Bottoms Flow would then be blocked back in. This "dump/fill cycle" would be repeated about every 24 hours, as illustrated in Figure 2.

Figure 2: Bottoms Flow Valve & Bottoms Accumulator Temperature

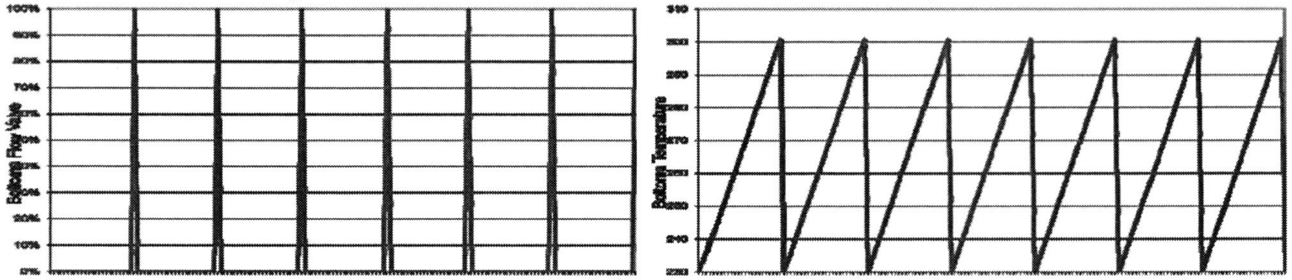

The solution to the stability and control issues was comprised of several elements:

- The first change was to the overheads control, swapping the MV-CV pairing based on an RGA as shown in Figure 3; a conceptual composition profile is also shown. Once tuned, the revised controls stabilized the mass and energy balance on the tower.
- Control Bottoms Flow based on Bottoms Temperature and tune it as an integrating process, based on the observed process response.
- The abandoned instrumentation on the Bottoms Flow stream and Accumulator were also upgraded as part of the controls' effort
- A Reflux/Feed controller was also provided for operator convenience to ensure sufficient heat was provided to the tower for the required fractionation.

Figure 3: Side Draw Tower Final Control Structure

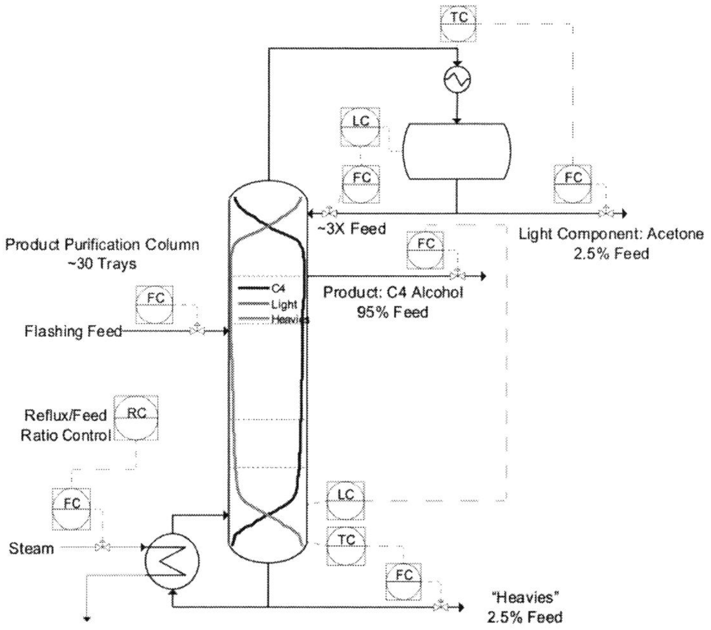

This initial example illustrates that even a simple integrated/intensified process, such as a side draw distillation column, can present challenges to robust and stable control. The analysis performed and the solutions performed indicated the need to understand the process fundamentals when proposing controls to these types of distillation units as "standard" solutions that are used on non-intensified process operations can lead to suboptimal solutions. Finally, this example also shows how a lack of proper instrument selection can create problems with unit operations in general.

Divided Wall Distillation Tower

This example covers a divided wall column (DWC) for solvent recovery. It is well known that DWCs are particularly challenging to design and to control. In this particular case, the level of complexity was raised further through the (apparent) reactivity of the mixture and the deep energetic integration of the column into the overall process.

In addition to the divided wall design, this design is further intensified by the fact that the solvent vapor supplies all the heat to the overall process. Refer to Figure 4. The solvent is Methanol which becomes diluted by water and contaminated by organic trace compounds. The removal of the organic compounds is of paramount importance, since they potentially contaminate the high purity product of the process. However, the objective of the DWC's design was not a sharp split. Remaining water in the Methanol up to a few percent is acceptable. Besides, the organic load of the water to be discharged should be as low as possible due to the cost for disposal and the loss of solvent through the side draw should be minimal due to the cost for solvent replacement.

Figure 4: Divided Wall Column

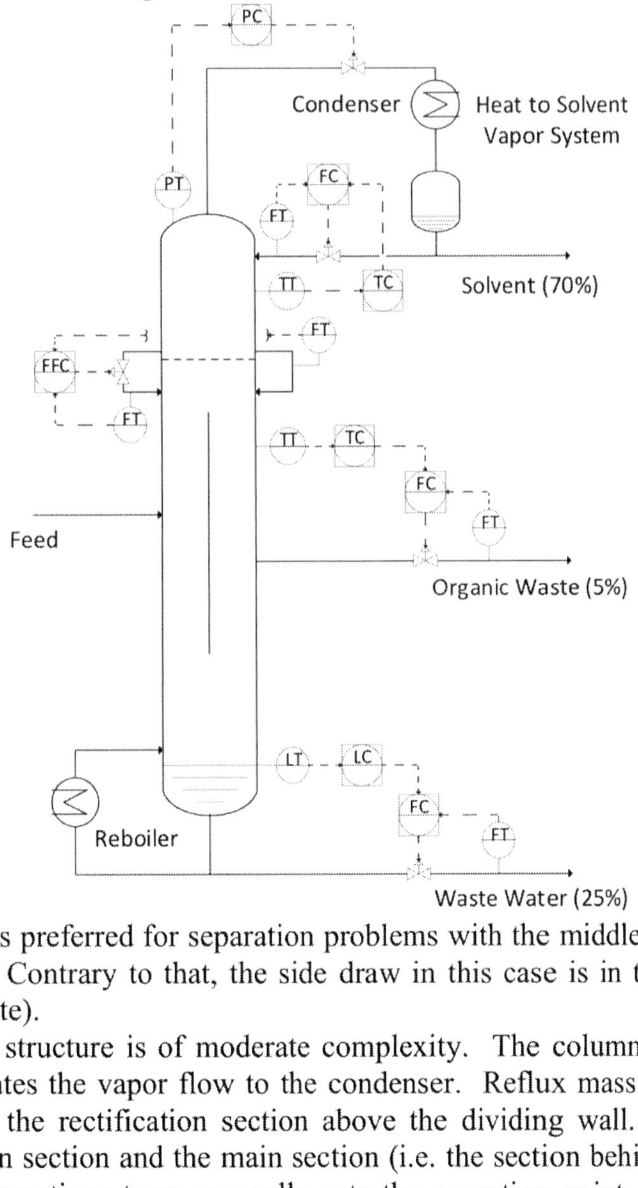

Generally, a DWC is preferred for separation problems with the middle boiler being present in relatively large quantities. Contrary to that, the side draw in this case is in the order of only a few percent (of the Feed flow rate).

The overall control structure is of moderate complexity. The column pressure is fixed by a control valve that manipulates the vapor flow to the condenser. Reflux mass flow is manipulated to control the temperature in the rectification section above the dividing wall. The liquid split ratio between the prefractionation section and the main section (i.e. the section behind the dividing wall) is adjusted by a valve; the operations team manually sets the operating point. The side draw rate is variable so as to fix a temperature reading between the side draw and the upper end of the dividing wall. The bottoms level was controlled by a mass flow controller.

The column was designed such that the overhead vapor would be sufficient to heat most other columns in the process. Hence, the column's reboiler was mostly running at full load. This further ensured that virtually no Methanol ended up in the bottoms and in the side draw.

During operation two significant problems arose:

(1) A light organic compound formed in the column with a boiling point similar to the process' main product. It left the column with the Methanol vapor (the solvent) and subsequently contaminated the product.

(2) At times, it was difficult to keep the column in the desired point of operation. Apparently, intermediate boiling compounds got trapped in the column and accumulated.

Observations and solutions:

(1) The light organic compound was one of the key trace compounds identified during process design. Hence, it was well understood how it formed and how its formation could be reversed. The column was further equipped with an acid dosing system to lower the pH in the prefractionation section. The originating compound, an aldehyde, was lower-boiling and as such easy to remove in the existing process set-up.

(2) Periodically, the accumulating intermediate boiling compounds were released with the Methanol to the solvent cycle. Since these compounds did not deteriorate the quality of the product, no immediate measures were taken. However, the built-up of these intermediate boiling compounds and the sub-optimal operation of the column impaired plant economics. And, as a side effect, the operating teams improved their skill sets by riding through these "periodic storms".

One of the options discussed was the installation of a second side-draw to periodically or continually purge the intermediate boiling compounds from the process and to monitor their built-up rate closely.

This example also illustrates both the great economic benefit and the problems that may arise from highly integrated designs. In case complex designs fail to perform as expected, it is incredibly helpful to have additional nozzles installed that enable the attachment and integration of additional measurement or sampling devices that help to better understand operational issues and facilitate their resolution.

Dairy Processes

Globalization together with the ever-increasing consumer demand for dairy products in recent decades have resulted in the creation of large-scale industrial dairy processing plants [18]. This is especially true for milk powder production where, for example, Fonterra New Zealand installed the world's largest dairy spray dryer which is able to convert 4-4.5 million liters of fresh liquid milk every day [19]. As a result of these large production volumes and high energy requirements, process intensification is a well-established practice in the modern dairy plant.

One area where process control been implemented to address complexities bought on by process intensification in diary production is in the water-removal part of the plant where multiple effect evaporators are common place; some facilities also incorporate Mechanical Vapor Recompression (MVR) to improve the energy efficiency of the operation which is a principle cost driver. There have been a few success stories that have been recorded in connection of controlling intensified evaporation process. In Karimi et al. [20], the authors had success in using inferential cascade control in operating a multi-effect falling film evaporator in a milk powder factory in Iran.

In comparison to the examples that have been discussed above (which are from the petrochemical industry), the dairy industry has some unique process understanding and process monitoring challenges. Firstly, to successfully control an intensified process it is necessary to understand the underlying process from a steady-state, as well as dynamics perspective. In the petrochemical industry, validated process simulations play a major role in this aspect. This is not the case in dairy industry where accurately modelling components in milk is difficult. A promising development in this arena is the creation of Pseudo-milk components for commercial process simulators which, in turn, allows for the modelling of entire dairy processes [21].

Similarly, the lack of monitoring methods available to accurately measure the quality of a production process further complicates the control of intensified (as well as normal) unit operations. Another promising development in this aspect has been the proposal of online monitoring tools such as hyper spectral imaging used in milk powder characterization [22].

In comparison to the petrochemical examples, the research results reported with dairy processes are laying the foundation on which intelligent process control systems can be developed for dairy unit operations. This is because the development of process simulation and monitoring tools lets one understand and identify the underlying phenomena as well as process dynamics which is of critical importance to successfully implement feasible process control solutions.

Lessons Learned

When analyzing these examples, it is clear that better understanding of process engineering fundamentals and process dynamics led to the development/design of better process control structures and/or unit operations designs. Once this understanding was established, the solutions proposed to successfully control the unit operations were relatively simple. This is an important finding as, in many situations, control engineers, as well as control academics, tend to suggest complex process control solutions to intensified distillation units which are both difficult and expensive to implement and also fail to be robust. Analyzing the success stories also allows for the following conclusions to be drawn.

I. In many situations the control in intensified processes is not well thought out during the process design phase
II. Considering the control of intensified processes at the design phase of a unit operation can be beneficial as this will guarantee the unit operation can be controlled at desired production targets once operational
III. Understanding the underlying process dynamics and phenomena prior is a prerequisite to developing stable and robust process control structures for intensified distillation units
IV. Process simulations and plant data visualization play a significant role in accurately identifying the underlying process dynamics and phenomena.

This paper has identified some of the criteria for the successful implementation of IIPs, but , clearly, there is a need for more research and, especially, industrial application of IIPs. Some suggested areas of research and industrial practice include:

- The integrated steady state and dynamic design of large scale plants;
- Further skills development and training of process design engineers;
- Development of new features in engineering and design tools;
- A systematic design and performance assessment & benchmarking; and, most importantly,
- The knowledge feedback and lessons learned from commissioned plants

References

1. Mansouri, S. S.; Sales-cruz, M.; Huusom, J. K.; Gani, R. Systematic Integrated Process Design and Control of Reactive Distillation Processes Involving Multi-elements. Chem. Eng. Res. Des. 2016, Submitted.
2. Stankiewicz, A. I.; Moulijn, J. A. Process Intensification: Transforming Chemical Engineering. Chem. Eng. Prog. 2000, 22–34.
3. Horiuchi, K.; Yanagimoto, K.; Kataoka, K.; Nakaiwa, M.; Iwakabe, K.; Matsuda, K. Energy Saving Characteristics of the Internally Heat Integrated Distillation Column (HIDiC) Pilot Plant for Multicomponent Petroleum Distillation. J. Chem. Eng. JAPAN 2008, 41, 771–778.

4. Roberge, D. M.; Zimmermann, B.; Rainone, F.; Gottsponer, M.; Eyholzer, M.; Kockmann, N. Microreactor Technology and Continuous Processes in the Fine Chemical and Pharmaceutical Industry: Is the Revolution Underway? Org. Process Res. Dev. 2008, 12, 905–910.
5. Heintz, S.; Börner, T.; Ringborg, R. H.; Rehn, G.; Grey, C.; Nordblad, M.; Krühne, U.; Gernaey, K. V.; Adlercreutz, P.; Woodley, J. M. Development of in situ product removal strategies in biocatalysis applying scaled-down unit operations. Biotechnol. Bioeng. 2017, 114, 600–609.
6. Lutze, P.; Gani, R.; Woodley, J. M. Process intensification: A perspective on process synthesis. Chem. Eng. Process. Process Intensif. 2010, 49, 547–558.
7. Nikačević, N. M.; Huesman, A. E. M.; Van den Hof, P. M. J.; Stankiewicz, A. I. Opportunities and challenges for process control in process intensification. Chem. Eng. Process. Process Intensif. 2012, 52, 1–15.
8. Schoenmakers, H. G.; Bessling, B. Reactive and catalytic distillation from an industrial perspective. Chem. Eng. Process. Process Intensif. 2003, 42, 145–155.
9. Ho, T. J.; Huang, C. T.; Lee, L. S.; Chen, C. T. Extended ponchon-savarit method for graphically analyzing and designing internally heat-integrated distillation columns. Ind. Eng. Chem. Res. 2010, 49, 350–358.
10. Harmsen, G. J. Reactive distillation: The front-runner of industrial process intensification. Chem. Eng. Process. Process Intensif. 2007, 46, 774–780.
11. Tuchlenski, A.; Beckmann, A.; Reusch, D.; Düssel, R.; Weidlich, U.; Janowsky, R. Reactive distillation — industrial applications, process design & scale-up. Chem. Eng. Sci. 2001, 56, 387–394.
12. Panda, D.; Kannan, A. Equilibrium and Rate Based Simulation of MTBE Reactive Distillation Column. 2014, 8, 998–1004.
13. Sneesby, M. G.; Tade, M. O.; Smith, T. N. Two-point control of a reactive distillation column for composition and conversion. J. Process Control 1999, 9, 19–31.
14. Pöpken, T.; Steinigeweg, S.; Gmehling, J. Synthesis and Hydrolysis of Methyl Acetate by Reactive Distillation Using Structured Catalytic Packings: Experiments and Simulation. Ind. Eng. Chem. Res. 2001, 40, 1566–1574.
15. Al-Arfaj, M.; Luyben, W. L. Comparison of Alternative Control Structures for an Ideal Two-Product Reactive Distillation Column. Ind. Eng. Chem. Res. 2000, 39, 3298–3307.
16. Georgiadis, M. C.; Schenk, M.; Pistikopoulos, E. N.; Gani, R. The interactions of design, control and operability in reactive distillation systems. Comput. Chem. Eng. 2002, 26, 735–746.
17. Chung, Y.-H.; Peng, T.-H.; Lee, H.-Y.; Chen, C.-L.; Chien, I.-L. Design and Control of Reactive Distillation System for Esterification of Levulinic Acid and n -Butanol. Ind. Eng. Chem. Res. 2015, 54, 3341–3354.
18. Munir, M. T.; Wilson, D. I.; Depree, N.; Boiarkina, I.; Prince-Pike, A.; Young, B. R. Real-time product release and process control challenges in the dairy milk powder industry. Curr. Opin. Food Sci. 2017, 17, 25–29.
19. GEA Building the World's largest dairy Spray Dryer http://www.gea.com/en/stories/building-the-worlds-largest-dairy-spray-dryer.jsp.
20. Karimi, M.; Jahanmiri, A.; Azarmi, M. Inferential cascade control of multi-effect falling-film evaporator. Food Control 2007, 18, 1036–1042.
21. Munir, M. T.; Zhang, Y.; Yu, W.; Wilson, D. I.; Young, B. R. Virtual milk for modelling and simulation of dairy processes. J. Dairy Sci. 2016, 99, 3380–3395.
22. Munir, M. T.; Wilson, D. I.; Yu, W.; Young, B. R. An evaluation of hyperspectral imaging for characterising milk powders. J. Food Eng. 2018, 221, 1–10.

591118 Pyrophoric Process Solids Mitigation

Abstract:

Pyrophoric materials pose significant process safety risks, particularly during equipment maintenance operations. Standard industrial practices to remove pyrophoric solids use mechanical or chemical methods, such as potassium permanganate treatment, which often require a process turndown. A novel pyrophoric solid has been observed in the commercial selective 1-hexene process. Characterization of this material by a variety of analytical techniques led to development of a deactivation method compatible with the process conditions. The pathway from laboratory discovery to commercial implementation will be discussed along with key safety learnings.

592114 Process and adsorbent for separating ethanol and associated oxygenates from a biofermentation system

Abstract:

Abstract for poster presentation or talk:

Abstract Title: Process and adsorbent for separating ethanol and associated oxygenates from a biofermentation system

Presenter Information: Deepak Sharma; Sr. Process Engineer, Engineering and Technology, Bayer AG

Patent number: US8658845B2

Patent Authors:

Deepak Sharma - Sr. Process Engineer, Bayer Ag, St. Louis; (Presenting Author)

Anil Oroskar – CTO, Orochem Technologies, Napervielle, IL;

David W. House – Scientist, Orochem Technologies, Napervielle, IL;

Alice M. Havill – Scientist, Lanzatech, New Zealand

The principles of process intensification if used on existing processes can lead to development and redesign of existing processes to become more sustainable. This presentation will discuss a process and an adsorbent for the separation of ethanol associated oxygenates from a dilute mixture of ethanol and associated oxygenates in water in the presence of organic compounds derived from a biofermentation process. After pretreatment, the separation is carried out in a simulated moving bed adsorption system employing a stationary phase adsorbent comprising fluorinated carbon or modified C18 silica gel selective for the adsorption of ethanol and associated oxygenates, such as 2,3-butanediol, with a mobile phase desorbent selected from the group consisting of methanol, ethanol, propanol, and methyl tertiary butyl ether. The process is useful for removing water from dilute aqueous mixtures of organic compounds comprising ethanol in dilute concentration in water and produced by fermentation, biomass extraction, biocatalytic, and enzymatic processes which are not economically recoverable by conventional distillation methods.

Keywords: Separation; Adsorption; SMB; Purification

Patent Assignee: Orochem Technologies LLC

592111 Pump safety, the undiscussed topic in chemical industry leading to several fatal incidents

Abstract:

Abstract Title: Pump safety, the undiscussed topic in chemical industry leading to several fatal incidents

Presenter Information: Deepak Sharma; Sr. Process Engineer, Engineering and Technology, Bayer AG

Pumps are one of the equipment's in the process industry that require a thorough review during HAZOP studies. The constant source of energy given to the pump, if not converted to kinetic energy to flow the fluid, can lead to heating of the pumped fluid. If the fluid keeps accumulating the energy provided by the motor, in a blocked-in scenario, it can lead to very high temperatures eventually leading to loss of containment of the pump and the pump can get damaged. There have been instances in the industry where fragments of damaged pumps have flown several feet away from the point of loss of containment leading to injuries. The poster will outline a scenario where a pump is run blocked in on suction and discharge for water and a chemical which has a low thermal decomposition temperature. Arc data is used to evaluate the onset temperature of thermal decomposition. The poster will outline safeguards that can be employed to avoid catastrophic loss of containment of the pump. Outlined below are safeguards used to avoid loss of containment of a pump in case of a blocked-in scenario:

- Install PSV on suction of pump
- Employ a low amp/power interlock on the pump to turn off the pump in case of blocked in situation
- Have a high-pressure interlock on the discharge line of the pump to direct flow back to the tank in case of pump deadhead
- Install an Interlock to turn off the pump in case the feed forward valves (automatic valves) close inadvertently

Pump deadhead scenarios are not only catastrophic when in flammable service there are several salts that have low thermal decomposition temperatures. Pumps in water service left blocked in could also blow catastrophically leading to reportable incidents. I feel this is an important topic to bring to the attention of students and professionals in the chemical industry. Poster will include pictures of actual installed pumps along with required safeguards in the field along with supporting P&ID depictions.

589440 New Kind of Linear Nanocomposite Applied in Improving Rheological Behavior of Crude Oil and Its Mechanism Studied through Molecular Simulation Method

Abstract:

In this work, a new kind of linear nanocomposite produced by carbon nanotube(CNT) as pour point depressant (C2-100) was developed by blending with MCNT nanoparticle and copolymer named Ethylene-vinyl acetate (EVA). Its effect of reducing yield stress and modifying the morphology of wax agglomerate were also evaluated. Besides, the synergetic effect of C2-100 nanocomposite with the magnetic field was also conducted. The mechanism of modifying rheological behavior of waxy crude oil was studied through molecular simulation method, which aims at understanding the interaction between this nanocomposite and wax molecule. The results indicated that C2-100 nanocomposite could reduce the yield stress of waxy crude oil more effectively than the copolymer (EVA) even at low concentration, as well as changing the morphology of wax crystals and agglomerate. Moreover, its effect of reducing yield stress becomes more obvious when the alternate magnetic field was applied. In addition, the results of molecular simulation also presented that the effect of improving rheological behavior of C2-100 nanocomposite is related to the adsorption behavior of wax molecules on the surface of MCNT nanoparticle, which is adsorbed with EVA molecules. In addition, the adsorption behavior of different fragments of EVA molecules can also influence the adsorption behavior of wax molecules on the surface of EVA-attached MCNT nanoparticle, which may influence the effect of C2-100 to improve the rheological behavior of waxy crude oil.

589683 Solar Energy Conversion and Noise Characterization Sensing in Building Integrated Solar Photovoltaic Devices with Ventilation

Abstract:

A study on solar energy conversion and noise characterization sensing in solar photovoltaic devices with ventilation is conducted. A building integrated parallel plate solar photovoltaic (PV) device was built with a pair of PV modules, a ventilated air cavity and an insulating back panel of plywood board filled with polystyrene installed in an outdoor test-room. Photovoltaic amplification was attained with power output from a potentiometer through rotation of its circular knob. A parallel plate PV device was studied for its
electrical parameters as Resistance-Capacitance (RC) electrical analog circuit. The effect of inductive and capacitive heating losses was considered in evaluating electrical characteristics of a PV device exposed to solar radiation. The characterization of noise interference, due to power difference of two intensities in a wave on PV Device is presented. Noise characterization of a wave is obtained depending on type of wave. The difference of two power intensities is due to transmission of light, sound, heat, electricity, fluid, and fire into a parallel plate PV device. Based on analysis of a solar PV device, a definition with measurement expression and unit for an electrical noise were coined. Standard definitions of noise sources, their measurement equations, their units and their origins under limiting reference conditions were also devised. A review for sensing noise of physical agents is conducted for sensing noise of physical agents (sun, light, sound, heat, fluid, electricity and fire). With integrated environmental control, these environmental parameters are to be characterized through various environmental monitoring sensors.

589945 Structural Characterization and Model Representation of Vitrinite-Rich India Coal By Using Advanced Analytical Techniques

Abstract:

The detailed investigation of the chemical structure of Karharbari coal (vitritine rich subbituminous coal from India) by utilizing advanced instrumental analytical techniques such as solid state ^{13}C nuclear magnetic resonance (^{13}C-NMR) spectroscopy, Fourier transform infrared (FT-IR) spectroscopy, X-ray diffraction (XRD), Flash-pyrolysis gas chromatography mass spectroscopy (Py-GC/MS), Laser desorption ionization time-of-flight mass spectroscopy (LD-TOF-MS), X-ray photoelectron spectroscopy (XPS), and high-resolution transmission electron microscopy (HRTEM). The Karharbari coal was an amorphous loosely organized lower rank coal with aromaticity 0.79. This was indicated by XRD results and also confirmed with ^{13}C-NMR results (aromaticity =0.84) and FTIR results (aromaticity = 0.82). The average molecular weight of the aromatic cluster is 507 amu by NMR result which is closer to the result obtained by HRTEM (MW = 530.46 amu). The presence of different heteroatoms like carbon, oxygen, nitrogen and sulphur with their functionalities, were determined by using a XPS technique. Different carbon/oxygen forms present in the Karharbari coal are (C-C) and (C-H) 68.53%, (C-O) 23.46%, (COO^-) 1.96%, (C=O) 6.05% and different nitrogen forms such as pyridine, pyrrolic, quaternary, oxidized and their distribution (mol %) are 19.36%, 45.68%, 31.25%, 3.71 respectively, and different sulphur forms were present i.e. corresponding to thiophenes, sulphones, sulphuric acid and sulphates with the molar content of 16.47%, 41.61%, 21.37%, 20.55% respectively. The construction of the individual molecules will be discussed in detail in the rest of this paper. The individual molecules were geometry-optimized and Materials Studio Amorphous Builder module was used to construct a three dimensional associated structure.

589479 Life Management of High Energy Piping with Highlights from ASME B31.1

Abstract:

High energy piping (HEP) systems – or as defined by the ASME B31.1 Power Piping Code, Covered Piping Systems (CPS) – are critical assets with respect to production facilities. A failure can have catastrophic consequences from a safety perspective and in terms of equipment loss and downtime. In fact, loss of high-pressure steam due to a failure in a main run of steam supply piping could shut an entire production unit down. Because of this and the requirements of the Code, HEP maintenance and inspection programs are established to ensure that the integrity of the systems is maintained in a safe and reliable condition throughout their lifecycle. There is no one-size-fits-all for the life management of these critical assets because there is a great variety in age, design, materials, and owner objectives. Life management of these assets is beneficial in that it provides a structured, systematic approach to creating and implementing a program including evaluating the current program (whether formal or informal), what the objectives of the program are (short and long term), what equipment is in place, and its history.

A HEP program put in place for overall life management is not synonymous with an inspection program, although inspections are an important part of an overall program. Inspections should be used in conjunction with analytical tools, real time monitoring, and laboratory examinations. The overall approach is life management – not inspection management – of the equipment.

This paper describes a HEP life management approach which includes identifying locations of interest, risk-ranking those locations, implementing a plan for determining the health of the highest risk locations, and taking actions based on the system health that has been determined. This paper also presents and discusses some of the requirements of the B31.1 code, which at the highest level states that each operating company shall develop operation and maintenance procedures for piping systems deemed necessary to ensure safe facility operations based on the provisions of the code, relevant industry experience, the operating company's experience (and knowledge of its facility), and the conditions under which the piping systems are operated.

590131 Optimization of C3MR LNG Liquefaction Process Using a Parallelized Simulator

Abstract:

Controlling the cost of energy-intensive processes such as C3MR (propane/mixed refrigerant) LNG liquefaction is a prime opportunity for optimization. Our work combines different C3MR models to develop an optimization that uses multiple independent variables, relaxed constraints, and flowsheet alterations to achieve an optimal operating cost. This work uses a rigorous model of the process in a process simulator rather than a reduced model of the process response. Use of a first-principles process simulator increases the accuracy of the model and therefore allows a more realistic optimization, at the cost of increased computing time. Parallelization (using multiple CPU cores) in the process simulator greatly increases the speed of the optimization calculations. CHEMCAD's optimization tools and multi-core technology make it possible to easily define the optimizations, run the optimizations efficiently, and generate global optima with a high degree of confidence.

590196 Modifying Fischer-Tropsch Reaction Sequence with Addition of Zeolite to Iron-Based Activated-Carbon Supported Catalyst

Abstract:

In addition to previously presented scientific work, modified reaction sequence of Fischer-Tropsch (FT) process was studied by addition of Zeolite ZSM-5 to an iron-based activated-carbon-supported catalyst. Main reactions involved due to FT catalyst are Fischer Tropsch Synthesis (FTS) and Water Gas Shift (WGS) reactions. Main reactions due to ZSM-5 are cracking, oligomerization, isomerization, cyclization, aromatization, conversion of alcohols to hydrocarbons, and coke formation. The waxy and linear structure FT products are converted to gasoline range, lighter hydrocarbons, iso-paraffins, and aromatics using ZSM-5. The catalyst ratio of 1:2, FT catalyst: ZSM-5, is used at 300 PSIG and 300°F. Separate-bed configuration, i.e. FT and ZSM-5 are separated in a reactor, arrangement is used for the reaction sequence analysis. Alcohols with carbon-chain from 3-5 produced by FTS are converted into hydrocarbons over ZSM-5. Waxy hydrocarbons with carbon-chain greater than 23 produced by FTS are cracked and isomerized to small n-paraffins and iso-paraffins using ZSM-5. The hydrocarbon with carbon chain from 7-10 produced by ZSM-5 cracking reactions are converted to aromatic products over ZSM-5. Produced gasoline to kerosene range n-paraffins are further cracked into smaller chain (1-5) iso- and n-paraffins. Some of the iso-paraffins are oligomerized and aromatized over ZSM-5 to form more aromatics. Further, multiple aromatic rings combine to undergo cyclization and dehydrogenation reactions over ZSM-5 to form coke. Water produced as a byproduct of FTS is combined with some of the carbon monoxide, which is the WGS, to form carbon dioxide and hydrogen.

590427 Lateral Retention Force Required for Sliding Saline Droplets over Silianized Silicon Surface

Abstract:

The lateral retention force required to slide a liquid drop on a surface is expected to increase with the time that the drop rests on the surface prior to sliding till it reaches a plateau. The molecular reorientation of the molecules at the periphery of the solid surface subjected to the unresolved vertical component of surface tension force causes this phenomenon. In this system, the surface tension of deionized water is increased by adding NaCl and is expected that the lateral retention force is increased proportionally with concentration. The retention force increased with increase in concentration reaching a maximum only to drop to a minimum with concentrations closer to saturation. That is, the lowest sliding force is required for both lower and higher concentrations of NaCl. This sudden decrease in observed lateral force is not a function of surface tension and the surface on which the drop slides. This phenomenon is associated with adhesion of salt to the surface and is supported by the drop sliding motion and depinning of the liquid from the solid.

590433 Study of Droplet Motion across Polymeric Substrates By Centrifugal Adhesion Balance (CAB)

Abstract:

On a given surface, the contour of the drop changes accordingly based on factors of the force applied, chemical structure of the fluid, drop size, and surface topography. With increasing force that is exerted on the drop, we determine the comprehensive trend for drop motion with respect to the controlled factors mentioned above. Drop motion is categorized into four phases and the point at where the initial displacement of receding line is observed is defined as the onset of motion. In this study, the onset of motion is affected by the factors mentioned above and aids in the determination for another value correlation known as Gs, known as Interfacial Modulus. The resistance to interaction towards the droplet as observed is quantified with respect to change in surface topography. This study therefore characterized drop motion and specifically onset motion based on controlled variables (factors). Work of adhesion is also quantified across different substrates to characterize the detachment of liquid from its substrate. All these parameters such as Gs, onset of motion, contact angles were analyzed with different polymeric substrates, drop sizes, and various fluids used.

590665 Population Balance Modeling and Spatially-Oriented Control for Crystal Size and Shape in Crystallization Processes

Abstract:

Crystallization is an important separation and purification process with relevant applications in the food, pharmaceutical, fine chemical and fertilizing industries. In these processes, modeling and process control for crystallization are focused on crystal size distribution (CSD) and shape of the crystals in order to make it possible to achieve the desired quality and purity of the product of interest. The control of crystal size and shape is of considerable importance for industrial crystallization processes, as crystal morphology has great influence on downstream steps and also from the point of view of product specification criteria. The consideration of crystal shape in modeling of crystallization processes is also important due to its impact on product effectiveness, such as the bioavailability and tablet stability in pharmaceuticals. Control of the crystallization process is can be facilitated with online measurements, which provide information to be used in the modeling of the process. However, if online measurements are not available, the strongly nonlinear dynamics of these processes makes difficult to establish relations between some inputs (e.g. temperature) and the final desirable characteristics of the crystal product. In this context, image analysis has been shown to be a suitable technique to monitoring the process [1].

In our present work, a population balance model-based scheme for controlling the size and shape of crystals is presented. The modeling was performed using a two-dimensional population balance (2D-PBM) in terms of the two characteristic dimensions of the crystals, describing the evolution of crystal size and shape distributions along the batch time. To identify the nucleation, growth and dissolution kinetics parameters using the 2D-PBM approach along the length and width directions, QICPIC-LIXELL's high speed dynamic image analysis equipment was used for the online measurement of the CSD, shape parameters and aspect ratio distribution of crystals during the experiments. In this way, the developed model allowed for the prediction of the crystal size and shape distributions during growth-dissolution cycles.

The developed framework was employed to obtain optimal temperature policies for controlling the crystal size and shape distributions that meet target criteria for size, morphology and productivity using the dynamic image analysis along the runs. Based on the concept of using measurements along batch crystallization processes as a trajectory in a phase space [2], a trajectory-endpoint control problem is presented with an oriented control scheme that is applied to the shape of the crystals. The scheme was used to control the batch cooling crystallization of potassium dihydrogen phosphate. The reported experimental data demonstrate the application of the control policies for temperature to produce crystals of the desired average shape and size in different batch times. Thus, using the first principle modeling coupled to the developed trajectory tracking control, the proposed framework has the potential to be applied to industrial crystallization processes for shape control and to be extended to polymorph control investigations.

References:

[1] Eisenschmidt, H.; Bajcinca, N.; Sundmacher, K. (2016) Optimal Control of Crystal Shapes in Batch Crystallization Experiments by Growth-Dissolution Cycles. *Crystal Growth & Design*, v. 16, p. 3297-2803.

[2] Griffin, D. J., Grover, M. A., Kawajiri, Y., Rousseau, R. W. (2015) Mass-count plots for crystal size control. *Chemical Engineering Science*, v. 137, p. 338-351.

590450 Auditing the Process Hazard Analysis - Key Characteristics to Scrutinize in a Good-Quality HAZOP Study and Supporting Effective HAZOP Study Updates

Abstract:

Although Process Safety Management (PSM), requiring good-quality Process Hazard Analyses, was promulgated in the United States in 1992, and HAZOP Study techniques date back to the mid-1970's, the HAZOP Study technique is still not uniformly applied in a high quality fashion. The objective of this paper is to provide experience-based tips on Process Hazard Analysis auditing, and some key characteristics that the auditor can scrutinize to assure quality implementation. Analogously, a similar set of characteristics that can facilitate future HAZOP Study updates can also contribute to a quality HAZOP Study that supports the spirit of Safety Management System objectives for the identification of hazards potentially impacting plant personnel, the community, and the environment. This paper will also review tips for structuring the HAZOP Study to facilitate later updates and revalidation.

590342 Development of High Performance Organic-Based Cathodes

Abstract:

For lithium-based batteries, structure/property relations in terms of cell voltage are desired for new organic-based cathode active materials. In this initial pursuit of structure/property relations, ab-initio calculations were performed on a few organic-based molecules to show the effect on the theoretical cell voltage by inclusion of nitrogen and/or oxygen. Preliminary test results are presented on a newly developed molecule and methods to potentially improve on the cell performance are given.

590705 Electrochemical Kinetics of Degradation of Graphite Anodes in Lithium Ion Batteries

Abstract:

Growth of the solid electrolyte interphase (SEI) is a major driver of capacity fade in LIBs. Despite its importance, the fundamental mechanisms remain unclear, primarily because of the complicated reaction pathways involved[1–3]. SEI growth can be both electrochemical and chemical in nature[4], and thus, it is a strong function of the potential and degree of lithiation of the electrode. We model the early-stage and long-term growth of SEI by accurately capturing the potential dependence of its formation kinetics as well as long term rate limiting steps, and validating it against the world's largest open source battery cycling data, generated in-house[5]. This is done using the Multiphase Porous Electrode Theory (MPET) framework[6] on graphite (phase separating) and carbon black (non phase separating) particles.

Lithium plating is another key degradation phenomenon that has been elusive, and it becomes important while trying to fast-charge batteries, i.e., 0% - 80% state-of-charge in 30 mins. We show that lithium plating is a key function of electrode morphology, phase-separation dynamics and potential. Phase-separation in graphite is modeled in the electrode using the Cahn-Hilliard Reaction framework described by Bazant[7]. We understand the electrochemistry of the onset of lithium plating with in-situ measurements connected to real time cell potential in a phase-separating electrode for the first time[8].

Results indicate that the peak SEI-forming currents are higher for higher driving currents. Also, we find that SEI only grows during electrode lithiation, i.e. the battery only degrades while being charged. We also find that onset of lithium plating is correctly captured only when phase separation in active material is accounted for. Further, the onset of plating is delayed on electrodes with a thick SEI layer – understanding SEI/plating coupling is integral to predicting fast charging manufacturing protocols for LIBs. This work holds promise for the predictive design of procedures[9] for manufacture and formation of LIBs.

[1] Cohen, Y. S.; Cohen, Y.; Aurbach, D. Micromorphological Studies of Lithium Electrodes in Alkyl Carbonate Solutions Using in Situ Atomic Force Microscopy. J. Phys. Chem. B 2000, 104, 12282–12291. https://doi.org/10.1021/jp002526b.

[2] Horstmann, B.; Single, F.; Latz, A. Review on Multi-Scale Models of Solid-Electrolyte Interphase Formation, Current Opinion in Electrochemistry 13, 62-69 2019.

[3] Nie, M.; Abraham, D. P.; Seo, D. M.; Chen, Y.; Bose, A.; Lucht, B. L. Role of Solution Structure in Solid Electrolyte Interphase Formation on Graphite with LiPF6 in Propylene Carbonate. J. Phys. Chem. C 2013, 117 (48), 25381–25389, https://doi.org/10.1021/jp409765w.

[4] Das, S.; Attia, P. M.; Chueh, W. C.; Bazant, M. Z. Electrochemical Kinetics of SEI Growth on Carbon Black: Part II. Modeling. J. Electrochem. Soc. 2019, 166 (4), E107–E118. https://doi.org/10.1149/2.0241904jes.

[5] Attia, P. M.; Das, S.; Harris, S. J.; Bazant, M. Z.; Chueh, W. C. Electrochemical Kinetics of SEI Growth on Carbon Black: Part I. Experiments. J. Electrochem. Soc. 2019, 166 (4), E97–E106. https://doi.org/10.1149/2.0231904jes.

[6] Smith, R. B.; Bazant, M. Z. Multiphase Porous Electrode Theory, J. Electrochem. Soc. 2017, 164 (11). https://doi.org/10.1149/2.0171711jes.

[7] Bazant, M. Z., Theory of chemical kinetics and charge transfer based on nonequilibrium thermodynamics, *Accounts of Chemical Research*, *46*(5), 1144–1160. https://doi.org/10.1021/ar300145c

[8] T. Gao, Y. Han, S. Das, T. Zhou, D. Fraggedakis, N. Nadkarni, C. N. Yeh, W. Chueh, J. Li, M.Z. Bazant, Interplay of lithium intercalation and plating on graphite using in-situ optical measurements, submitted.

[9] Huang, W.; Attia, P. M.; Wang, H.; Renfrew, S. E.; Jin, N.; Das, S.; Zhang, Z.; Boyle, D. T.; Li, Y.; Bazant, M. Z.; McCloskey, B. D.; Chueh, W. C. and Cui, Y.; Nano Letters 2019 19 (8), 5140-5148. DOI: 10.1021/acs.nanolett.9b01515

589787 Model Development of Solid Oxide Fuel Cell Thermal Performance Using Artificial Neural Network

Abstract:

NASA's Johnson Space Center has recently begun efforts to eventually integrate air-independent Solid Oxide Fuel Cell systems, with landers that can be propelled by LOX-CH4, for long duration missions. Using landers that utilize such propellants, provides the opportunity to use such systems as a power option, especially since they will be able to process methane into a reactant through fuel reformation. To ensure fuel reformation in the systems, Steam Methane Reformation (SMR) will be employed. Various lead-up activities, such as hardware testing and computational modelling, have been initiated to assist with this developmental effort including the development of dynamic empirical models using artificial neural network. Such a modeling approach has shown to be accurate (R^2>0.9) to predict the thermal and electrical performance for SOFC systems. Hence, such models could be used for system optimization and control design.

Impact of Spacer Nature on Rheological Properties of Cationic Gemini Surfactant

Abstract ID: 589650

Shams Kalam[1], Muhammad Shahzad Kamal[2], Shirish Patil[1] and S. M. Shakil Hussain[2]

(1) KFUPM, Dhahran, Saudi Arabia

(2) Center for Integrative Petroleum Research, King Fahd University of Petroleum & Minerals, Dhahran, Saudi Arabia

Surfactants play a vital role in Chemical Enhanced Oil Recovery (cEOR) techniques. Gemini surfactants are a unique style of surfactants having more water-soluble hydrophilic head and oil-soluble hydrophobic tail units. Both groups are connected by a spacer at or near to the head groups. Polymers are added into the surfactants to increase the sweep efficiency of the flooding processes. This paper presents the rheological properties of Surfactant Polymer (SP) solutions in which two recently synthesized cationic gemini surfactants with different spacer nature are used.

The rheological properties of novel SP solutions were determined using Discovery hybrid rheometer (DHR-3) from TA Instrument. The geometric configuration set for this work was concentric cylinder. All experiments were carried out at 80 °C, while the shear rate for steady shear viscosity measurements ranges from 0.001 to 1000 s^{-1}. 0.25 weight % of a polymer FO 4290 SSH was used with different concentrations of newly synthesized gemini surfactants for Rheological measurements. Two hours magnetic stirring and twenty hours retention time were provided to make uniform surfactant polymer solutions.

Both surfactants exhibited good thermal stability in seawater with no precipitation. Shear viscosity and storage modulus were measured as a function of shear rate and angular frequency respectively at different surfactant concentrations ranging from 0 to 0.5 mM, where 0 mM shows pure polymer. We discovered that counter ions also influence the rheological properties significantly. Research outcomes revealed that the novel SP solution with a benzene ring and chlorine counterion produces better performance in comparison with benzene ring and bromine counterion. Moreover, the effect enhances if we increase one more benzene ring in the spacer.

This study reports two recently synthesized SP solutions for cEOR processes in carbonate reservoirs. It presents the effect of counterion on rheological properties and improves the understanding of tuning the surfactant structure in making SP solutions to increase the flooding sweep efficiency.

590257 Real Time Energy Management Systems Implementation for Continually Optimized Performance

Abstract:

It is well known that energy management is key for the optimal operation of refining and petrochemical sites. Industrial energy systems are constituted by complex utilities networks that are interconnected and have complex interactions that in order to achieve it best performance need to comply with increasing environmental rules, push against operating constraints and adapt to altering power prices and rapidly changing equipment availability.

After the initial installation, management and continuous improvement of these systems present challenges not only from a technological but also from an organizational point of view.

The implementation of an advanced digital Energy Management System (EMS) which includes capabilities for multi-period optimized scheduling, real time optimization and monitoring, deployed either as cloud-based or on-premise solutions that could be acting on open or closed loop is known to provides large benefits.

Critical is the sustainability of the EMS in the ever-evolving energy supply and demand of refining and petrochemical sites. Commercial, asset and organizational factors all change and the EMS needs to adapt to the developing environment of each enterprise meaning the type of implementation, including appropriate management structure, culture, best practices and Key Performance Indicators (KPIs) needs to be continually improved.

The EMS implementation at the large and complex Eastman site in Longview, TX, will be presented as an example. It will provide insight on the initial successes achieved in real time energy management for both, cost optimization and energy related KPIs performance tracking. The case study evolves to discuss how technical asset improvements, once replicated in the EMS, provided alternative operating modes, large additional benefits and sustained optimal

Andrew Lammers, NN2

Eastman Longview

Pablo Montagna*, Cristian Rojas

KBC Advanced Technologies, Inc

589818 Analysis of Fuel Cell System through Project Based Learning for Application of Classical Thermal Fluid Concepts in Mechanical Engineering Laboratory Course

Abstract:

A hands-on project involving analysis of air noncircular channels in Fuel Cell to measure the pressure was introduced in a thermal fluid laboratory course. Although the students studied fluid mechanics as a theoretical course, this project gave them the opportunity to build an experimental setup from scratch to help showcase their hands-on skills and to apply their understanding of relevant concepts of fluid sciences. The results of the project included a successful test of a conceptual procedure, designed to measure and calculate pressure drop and flow rates across the anode and cathode sides of a fuel cell. The students enjoyed completing and learning from the project. They also provided very useful feedback on the project by incorporating a functional experimental system for the future laboratory project. This project based learning opportunity allowed students to build their confidence in what they are learning as well as provide a glimpse of what the real world may hold.

589438 Measured Versus Calculated Sour Gas Properties

Abstract:

Measured versus calculated sour gas properties
Abstract:
Recent Environmental agencies have restricted standards for sulfur emission from gas chemical processing. Sour gas can extremely impact the delivery and processing by corroding the equipment. Gas properties such as compressibility, density and viscosity are a major factor controlling prediction of future production pipelines, and production tubing and pipelines in gas industry.
Sour gas effect on all gas industry aspects usually gas properties are measured in laboratories but occasionally experimental values are not available due to complexity and time consuming hence a reliable method would be proposed.
We are assessing the accuracy of the industry standard methods in calculating sour gas properties, compressibility factor using the following correlations:
Drunchuck - Abukkasem, Hall – Yarbarough, Drunchuck- Purvis- Robinson, Londono et-al, Alanazi et-al and Azizi et-al.
Viscosity using empirical correlations such as Carr-Kobayashi Burrows, Lea-Gonzalez –Eakin and Dean- Stiel.
Density using equations of state Redlich Kwang and Soave -Redlick - Kwang.
This study shows that the best correlation for calculating compressibility factor and density of high concentration sour gas is Londono-et al, all other considered methods are not accurate. All other methods investigated include high error especially for high H2S concentration.
However the error in estimation of Z factor for most of the considered methods has great impact on estimated sour gas density. Therefore we recommend that sour gas properties be measured experimentally at high pressure and high temperature conditions.

587967 The Study of a S-CO2 Oxy-Fuel Combustion Power Plant for Low Grade Coals

Abstract:

Recently, it is very important to have reliable solutions that can maximize power generation efficiency of current power generation processes and decrease emissions of CO_2 and other pollutants because there are increasing concerns about the global warming and the depletion of fossil fuels. Among the available carbon capture technologies and power generation technologies, a S-CO_2 (supercritical carbon dioxide) cycle based oxy-fuel combustion process appears to be a promising mid-term solution for the production of electricity from fossil fuels. Since supercritical carbon dioxide (S-CO_2) involves a rapid density change at small temperature and pressure ranges, the S-CO_2 cycle can extract large amounts of energy from relatively small equipment and it is possible to minimize or simplify a power generation block and to obtain higher thermal efficiency than existing steam cycles used in power generation plants. Therefore, an indirect S-CO_2 oxy-fuel combustion power plants for low grade coals was designed and analyzed by using ASPEN-PLUS. Also, the operating characteristics of the indirect S-CO_2 Oxy-fuel combustion power plant were compared with those of conventional air combustion power plants.

590099 Characterization and Performance Assessment of Oil Entrapment By Commercially Available Surfactants

Abstract:

Oil spills occur frequently and are a threat to the health and safety of human society and the well-being of the environment and ecosystem. Remediating oil spills is an ongoing challenge. Surfactant based dispersing agents, such as the Corexit® product line, are widely used for remediation of oil spills mainly through direct application on top of the oil slick. This work encompasses the investigation of variables that influence the emulsification process and therefore the encapsulation of oil within the surfactant micelles. Of central importance are the conditions under which oil entrapment in emulsions even occurs. A central finding is that there appears to be a threshold shear rate for emulsifying oil with surfactants, below which no emulsification occurs. Different schemes for applying shear stress in this work included simply stirring, using a high energy colloid mill, and flowing the mixture through a high shear capillary. Also, the sequence in which the oil/surfactant/water (O/S/W) components are mixed impinges strongly on whether and how much oil entrapment occurs. To try to correlate shear stress and velocity profiles throughout the sample volume with entrapment efficiency and kinetic stability of these O/S/W systems, Computational fluid dynamics (CFD) simulations were conducted. Another important factor is the ionic strength of the aqueous medium, which greatly impacts kinetics of these emulsions, but with a smaller impact on the threshold at which any emulsion is obtained at all. Surfactants tested include Corexit® EC9500A, sodium dodecyl sulfate (SDS), Triton X-100, and cetyltrimethyl ammonium bromide (CTAB). The quantification of oil entrapment was performed through either thermal gravimetric analysis (TGA) or differential scanning calorimetry (DSC).

Designing a Native Model for Assessment of the Effectiveness of Advertising

Sorush Niknamian
PhD, military Medicine Department, Liberty University, USA
Email: so.niknamian@gmail.com

Abstract

Reviewing and evaluating the effectiveness of advertising can be used as a criterion to convert weaknesses into strengths and budgeting. Also, Identification of the factors influencing the effectiveness of advertising can lead to more professional design and implementation. In the banking industry, with the increasing number of banks and increased right of customers to choose, the role of advertising has become increasingly important to succeed in this field of competition; so it should be said that advertising can make the bank close to its goals when it is sufficiently effective. The current research aimed to identify and rank the effective factors on the design of a native model for evaluation of the effectiveness of advertising of the Eghtesad Novin bank in Iran. The statistical population of this research is all managers and experts of the Eghtesad Novin bank. According to the research literature, the criteria and sub-criteria affecting the effectiveness of advertising have been extracted. By using the fuzzy Delphi technique and based on the experts' opinion, the final indices were extracted. Then, using the fuzzy AHP method, the rank of criteria has been addressed. The derived criteria are demographic characteristics, individual emotions, and attitudes of the customer, apparent features of the message, message content, market characteristics, cultural, economic features, design and nature of the service, advertising methods. The results of the ranking showed that the methods of advertisement, the design, and nature of the service, the apparent characteristics of the message, the emotions and attitudes of the customer, the demographic characteristics, the content of the message, economic, market characteristics, were placed in the first to ninth ranks, respectively.

Keywords: banking services, banking services marketing, advertising, advertising effectiveness,

Introduction:

Advertising is one of the promotional mix elements of marketing mixes, which play an important role in attracting customers. Advertising is a bridge between the organization and customers and can introduce the organization and its products to customers, inform consumers about the goods and services of the organization, convince them about the benefits of the product, and ensure them to have the best choice even after the purchase when the customer is uncertain (Buck, 2016: 172).

Given the increasing competition among the brands in the marketing sector, advertising is a necessary tool to introduce the characteristics of goods and services to customers. Leong et al. (1999) pointed out the fact that consumers overcome advertising information, and many of them do not remember the advertisements that they are exposed to, and it is necessary to consider the advertising and its attractive methods to attract customers' attention. The ability to attract customers' attention is one of the main indicators of advertising effectiveness that should be taken into consideration. The attention of customers is a key component of the overall effectiveness of advertising (Pileliene et al., 2017: 2). Any professional marketing without advertising will not be effective. Despite relatively high advertising costs, advertising is mentioned as an investment rather than the cost. Since many advertisements cause that the advertising companies do not benefit their advertising because of the wrong way of advertising, the competitive companies benefit from the advertising of that company (Hartnet, 2016: 23)

Advertising should be in line with the goals and plans, as well as the strategies of the organization. Therefore, we should see to what extent the advertising has met such goals. Allocating funds without evaluating the effectiveness of past advertising is the waste of resources of the organization. By evaluating the effectiveness of advertising, changes can be made to the form, the content of messages, type of media, budget, communication channels, and even the time and conditions for conducting advertising to create more useful and effective advertising than the past (Yazgirudo-Yasteh, 2015: 359). In the current competitive world, the importance of advertising is increasing. As a result, businesses have devoted a great deal of money to customers' attraction, but it is important to see whether the company advertising has been designed and implemented in such a way to meet the goals of the company. This issue has been often neglected in our country. (Kim, 2014: 2107).

The amazing effect of advertising on the lives of individuals requires that this scientific, social and communicative branch be carefully studied and advertising is made with a scientific, artistic, specialized and expert view. Professional advertisement can be a sharp sword for sales promotion, both inside and outside the country, and otherwise, it is a waste of time, money and energy. A professional advertising lever can, with the help of new technology, penetrate humans, and cause more desire for consumption (Arribeh, 2016: 463). Today, banks have turned their attention to the widespread use of mass media advertising, given the highly competitive structure governing on them. Also, given the relationship among a wide range of businesses of organizations and households and banks through various deposit accounts, loans, and other financial services, banks can benefit from advertising in a proper way to meet their goals, namely the introduction of their services, the communication with customers, their effectiveness on them in choosing a bank to do their banking business and ultimately attracting them. Unfortunately, in most banks, despite the massive advertising and the allocation of large amounts of energy and budget, the effectiveness of such advertisements is less evaluated. Since advertising is one-way, it is necessary to receive feedback from the customers (Rabiei, 2011: 19).

Regarding marketing, the effectiveness of advertising depends on the wide range of coverage of various consumers and the cost of advertising campaigns. Economists have defined the effectiveness of advertising as a change in the profit derived from promotional activities. Therefore, the effectiveness of bank advertising is the potential quality of information reported to consumers, and in promotional campaigns, it aims at creating a bank image and informing banking products and services, as well as obtaining a specific economic result before and after the transmission of the message (Denis et al., 2012: 85).

Eghtesad Novin Bank (EN bank), which is one of the oldest banks in Iran, due to the addition of new banks, should use new tools to protect and even increase its market share for the assessment of the effectiveness of its advertising. For this purpose and given that the bank uses various advertisements, it is necessary to design a model to evaluate the effectiveness of the bank advertising; therefore, the present study seeks to answer the following questions:

- What indicators are used to assess the bank native advertising model?

- How are the indicators for bank advertising evaluation weighed and ranked?

In this research, we have tried to identify the criteria using the Fuzzy Delphi process; then we will use the Fuzzy AHP method to rank the identified criteria. In the next section of the research, the literature has been studied. Then the research design has been expressed, and in the last section the conclusion has been presented.

1-Research literature:

1-1-Marketing of banking services

The growing trend in service activities has led to the emergence of new markets and large service companies. The complexity of service industries, such as information systems, management consultancy, and logistics, has increased, as service providers have entered the market with a variety of business backgrounds and capabilities to provide a broader range of services. Facing this complexity requires knowledge and skills to understand the services. The definition of the service includes all activities that have been done (for example, writing document specifications) (Selviaridis et al., 2013: 1398).

In the industrial literature, there are various ways to define the service and product requirements. Van Weele (2010) has distinguished technical and applied definitions from each other. The first requirement expresses the exact technical specifications of the goods with the required services and inputs while it focuses on performance demands and demand-oriented performance orientations. In this regard, Axelsson and Wynstra (2002) have proposed four methods for definition of services: input definitions are based on the sources and capabilities of the service provider while the process-oriented approach focuses on how services are presented regarding activities and processes. Output-oriented definitions focus on service capabilities and performance levels, as well as the interaction between the suppliers and recipients. Ultimately, they focus on stress-oriented results in the economic value of services (Selviaridis et al., 2013: 1399).

The financial sector forms an essential part of the service sector. In this regard, banks have used marketing to improve the strategic performance of banks. In particular, since the 1980s, this trend has been confirmed and criticized. Marketing researchers and experts are interested in researching the various aspects of marketing in banking. The quality of services is essential for customers who use banking services, and the provision of accurate and high-quality services leads to the survival of banks in the competitive market. Improving the quality of services will increase the access and response to the demands of the bank's customers, which will make the bank more distinct than other competitors. Therefore, measuring the quality of services in the banking industry is essential. Providing quality services is a hard and difficult task, in this regard, the features of services in general and banking services, in particular, should be explained. Banking services require a high degree of appropriate communication between service providers and customers. If services are not provided properly, they will be at high risk.

About the five features that distinguish them from material services (intangibility, non-allocation of materials for production, diversification, non-corruption, and lack of ownership), financial services have two other specific features, that have been firstly developed by McKechnie (1997). These characteristics include trusteeship and two-way information. These special features mean that the evaluation of services is more difficult than material products, and they not only create a challenge in the development of marketing activities, but also they are the source of the problem in providing an optimal level of quality services. Banking services require a high level of communication between service providers and customers, as well as a high degree of customer participation in the process of providing services. Angurr et al. (1999) emphasized that due to the high standardization and the development of technologies in the banking sector, human factor has been considered as the most important one in making distinctions among competitors. The expertise and motivation of the employees, as well as the level of customer involvement in the provision of services, significantly affect the quality of services. Due to the wide range of banking activities, attention should be paid to cultural and social changes that affect the attitudes of customers (Durdana et al., 2015: 100).

Marketing of banking services has three components as follows: if any of them were properly executed, banking services would change the country.

1- Market study: Identification of the banking services market system, including the status of current banking services and their weaknesses, strengths, expectations, demands, ideas and opinions of customers, the conditions of environmental instruments affecting the market system, in particular government policies and governing laws, and even global changes in the banking system and successful models in this system. Indeed, market study is the foundation of the evolution of the banking system, because market study (which is the result of constant communication with the market and conducting marketing studies and research in various forms) leads to comprehensive, up-to-date, relevant and necessary information for decision-makers and bank managers (Munsung Rheea & Satish Mehrab, 2006: 508).

2. Market creation: It involves the use of a variety of simple and complex marketing techniques and competitive strategies, as well as the growth, development and market creation in the banking system; it needs to understand the internal customers (employees) and external

customers, to communicate with them and meet their needs and demands. Market creation in the competitive environment requires appropriate strategies (Moch, 2013: 2908).

3. Market protection: One of the most important tasks of marketing in the banking system of the country is to maintain market share, a stable relationship with customers, and to convert ordinary customers to loyal and supportive ones. The more intense the competition, the more valuable the loyal customers will be, as marketing research in the world has shown that the cost of finding a new customer in firms is 3 to 33 times higher than the cost of maintaining current customers regarding the type of products and services. Therefore, market protection prevents the costly advertising, and attraction of new customers and provides a kind of confidence and serenity for market survival, especially in competitive conditions (Fredriksson & Moro, 2014: 69).

1-2-Effectiveness of bank advertising

One of the indicators of success in any organization is the extent to which the organization achieves its goals. Here, the success of any organization is determined by its own goals. This approach has been called effectiveness in the management literature. The effectiveness is the degree to which the organization achieves its goals. Goals must first be defined to measure effectiveness in any organization. Then, plans are set up to fulfill these goals, and thus the plans will be implemented. In the end, the results are compared with predetermined goals (Manjegani et al., 2015: 124).

About 100 years ago, William Heketh statement, "Half of the advertisements were wasted, we just do not know which half it is" attracted everyone to the accountability of advertising. What matters to advertisers is that how much the amount of money spent on advertising has returned? Has the cost spent on making and broadcasting the advertising been effective? Has the effect of advertising been strong or weak? Knowing the answer to these questions is important to the advertiser because today companies pay much money to make and broadcast advertising. According to Jack Trout, the money we use to advertise is very much. Every second of the World War II was $ 9,000, and Vietnam cost $ 22,000 per minute. But in the American Football Final, every minute of advertising cost $ 2 million, so the advertiser has the right to know about the effectiveness of his advertising (Aksoy, Atilla, 2005, Yeni Reklamcilik, Istanbul Bilgi Universitesi, 95). It is very difficult to measure the impact of long-term advertising on many occasions. Also, academically in the world of advertising, it is impossible to create a comprehensive theory that can be used to determine the effectiveness of advertisements on people at different levels under different circumstances. The effectiveness of media is its capacity to create desired results, or special conditions in a particular position and content, that is, attaining the desired communication goals (Wery, 2012)

Evaluating the effectiveness of advertising indicates how much advertising has succeeded in achieving the goals. The evaluation of the effectiveness of advertising acts as a feedback, informs us about the weaknesses and strengths of our advertising by which we find our shortcomings in the advertising. The marketer may choose an appropriate stage regarding product life or brand position, and use it to define his advertising goals and ultimately evaluate its effectiveness (Eckler & Bolis, 2011: 2).

The effectiveness of advertising is a function of its content (message), the execution (how it is transmitted), and repetition (the number of times each consumer sees the message) (Cutler, 1997, quoted by Moorthy & Hawkins, 2005). Some of the effectiveness criteria are the call and recognition of the brand name (Park, 2006), awareness, willingness and intention to purchase (Decrop, 2007), attitude towards the ad, attitude toward the brand name and intent to purchase (Muda et al., 2014)

The control of the communication element has a direct effect on customer perception of different dimensions of the brand name, such as service quality and loyalty. It means that the elements of communication, advertising and promotion must be carefully implemented and controlled (Suvittawat, 2015). Informative advertising is one of the factors that shape the behavior of the bank customers. Studies conducted both on attracting, retaining and loyalty of customers, show the need to examine the impact of advertising on attracting and loyalty of customers (Ioannou, Boukas and Skoufari, 2014: 25).

An ideal model of world-class advertising can: 1. transfer a consistent meaning to the consumer all over the world. 2. Transfer positive global capacity for the brand. That is, the results of advertising among different peoples should produce a positive effect and transfer high-quality signals (Strebinger et al., 2018: 410).

Advertisers want to prepare consumers to like goods or services, but sometimes it harms them through unwanted and intrusive advertising. It leads to the creation of a series of negative emotions in advertising. The purpose of ads is to create positive feelings in consumers about advertised goods and services (Bell & Buchner, 2018: 1).

On the other hand, research shows that commercial advertising can be positively evaluated and considered, and this positive evaluation leads to a positive evaluation of the brand with the product. However, advertising can have negative effects that lead to the formation of negative attitudes towards that product and service. Therefore, by increasing the volume of messages, it can have positive, negative or neutral effects. Therefore, marketers should evaluate advertising messages and their behavioral outcomes. With this regard, the effectiveness of advertising, based on a hierarchical approach, should be fully investigated (Tafani et al., 2018: 125).

There are various factors which influence effectiveness of advertisement like media, time, market life stage, advertisement life stage (new or old advertisement) and also advertisement creative signs (like content and administrative elements) (Talis et.al, 2005). Also using popular cartoon personages (Ranjbarian, 2009), size (Rousat, Venous & Ebrahimi, 2004, Baltas, 2003), color (Park, 2008, Moriarty, Mitchell, Wells, 2009, Cheng & Kao, 2011), complexity or simplicity of the advertisement (Morrison & Dainoff, 1972, Di Res, 2008), using irony factor (Mukherjee, Hober, 2004), using famous artists (Rousta, Venous, Ebrahimi , 2004) and consumer features (Matrin, Phaymi, Agee, 2002, Park, 2008) are among these variables. Advertisement repetition is also a factor which helps in reminding its content to the audiences (Rousta, Venous, Ebrahimi, 2004). One of considered factors in the debate of evaluating effectiveness of advertisement is the attitude towards advertisement. Attitude towards the advertisement includes public interest or lack of interest in an advertising stimulus in encountering a special topic (Moaen & Minor, 2009). Mental imagery (B one & Elen, 1992), content transparence and emotions called through the advertisement are among variables influencing the attitude towards the advertisement. Some of these factors may lead to strong impacts while some of them may create weak impacts or don't include any impact. Awareness of the impact of each of these factors makes a considerable help in making management decisions (Talis et.al, 2005).

Fotini et.al (2009) provided a model of effective advertisements for advertising websites of big companies. The results of the study showed that measurements, activities, priorities and intentions of people have a considerable impact on evaluation and qualitative factors of relation-oriented marketing have a great impact on advertisements effectiveness indicators. The extracted criteria include:

- Advertisement policies: product information, company information, product general display, consistency with traditional advertisement

- Medium features: message features, information structure, responding time, facilitating research, beauty, interactions, communications in real-time, amusement, information control

- Relation- oriented marketing policy: communications, feedback, supporting customers

Today, organizations for competition in the business world and selling their goods and services require purposeful and efficient marketing activities. One of topics which has a very important role in the process of marketing of an organization and assign a high volume of budget to it is advertisement. Advertisement causes some changes in the knowledge, attitude and behavior level of audiences, so attention to evaluating advertisement effectiveness is a necessity for the organizations that consider advertisement as requirement of continuous life, growth and development. By assessing advertisement effectiveness and investigating consistency with the organization objectives and its expected impacts, some changes may be created in its dimensions, content and methods so that a more effective and useful advertising is performed. Regarding the research literature and internal and external background which has been stated, here a general categorization of the model proposed criteria has been provided in the following table:

Table 1- summation and extracted variables

scholar	sub standard	criterion
) Dargi & Mohammadian (2010), Ashouri (2016), Mansouri Moayed (2005), Aghapour (2016), Mirzadh & Mehrmanesh (2016), Keramforn (2011), Mac Ki- Nesbiat et.al (2011), Rosemary et.al (2012), Wang (2015), Maryion et.al(2017)	awareness of sexuality differences	individual factors
Dargi & Mohammadian (2010), Ashouri (2016), Samsam Shariat Keramforn (2011)(2007), Mirzadeh & Mehrmanesh (2016), Mac Ki- Nesbiat et.al (2011), Rosemary et.al (2012), Wang (2015), Maryion et.al(2017)	attention to age requirements	
Dargi & Mohammadian (2010), Dehdashti Shahrokh (2009)	various consumers (permanent, ordinary and case consumers)	

Khashizadeh (2010), Danaei & Dasar (2016), Pillay et.al (2012), Marion et.al (2017)	advertisement richness	
Danaei & Dasar (2016), Dargi & Mohammadian (2010), Pillay et.al (2012), Olson et.al (2012)	advertisement accuracy	
Ashouri (2016), Rabiei (2011), Taeini Harari (2014)	positive mentality of the organization	
Dargi & Mohammadian (2010), Shahmohammadi (2011), Olson et.al (2012)	honesty feeling in advertisement	
Fakhari et.al (2012)	threatening feeling	
Ashourizadeh (2016), Hamidizadeh (2013)	attitude towards brand	
madhoushi (2013), Olson et.al (2012)	being convinced	
Ahmadi (2013), Mac Ki – Nesbiat et.al (2011), Olson et.al (2012)	convicting audiences	
Ashouri (2016), Hosivar (2014), Olson et.al (2012)	mental imagery	
Siadat & Sakhi Jourashk (2014)	positive feeling	
Navaei et.al (2015), Rozmari et.al (2012)	audience being active	
Mosavi Esmaeilzadeh (2015), Olson et.al (2012)	knowledge level	
Dargi & Mohammadian (2010), Ashouri (2016), Shafiei Nikabadi & Zarei (2017)	attitude towards advertisement	
Dargi & Mohammadian (2010), Hosseinzadeh et.al (2016), Sanei et.al (2016)	familiarity with methods and forms of advertisement	message
Ashouri (2016), Olson et.al (2012), Alan (2012), Patric Ro (2014)	attention to reduction of expectation time of the user for observing advertisement	
Danaei & Dasar (2016), Forentebury & Mac Gldric (2011)	beautiful and appropriate	

	combination of colors	
Aghapuor (2016), Sanei et.al (2016)	site and location of advertisement placement	
Hassanzadeh & Danaei (2016), Sanei et.al (2016)	access to sufficient and required information	
Dargi & Mohammadian (2010), Darzban (2016)	starting the advertisement with a challenge	
Navaei et.al (2015), Salimi Fard & Pourdehghan (2016), Bargar et.al (2016)	starting the advertisement with a statement and explanation	
Dargi & Mohammadian (2010), Mirzadeh & Mehrmanesh	using helping words and terms for decision making	
Dargi & Mohammadian (2010), Daneshvar et.al (2016), Alan (2012)	using stimuli with diversions more than usual and expected extent	
Sanei et.al (2016)	using images of women and children	
Sohrabi Ranani (2016), Fortenbury & Mac Goldric (2011)	creativity in the message	
Shafei Nikabadi & Zarei (2017), Denis et.al (2012)	the text of advertisement message being a predicate sentence	

Khashizadeh (2010), Dargi & Mohammadian (2010), Sohrabi Ranani (2016), Patric Ro (2014), Abdolsalam et.al (2016)	continuation and repetition of message	
Daneshfar & Nabizadeh (2015), Fortenbury & Mac Goldric (2011)	using trustable and famous people	
Salimifard & Pourdehghan (2016)	common dialects and terms	
Daneshfar & Nabizadeh(2015), Forentnburi & Mac Goldric (2011)	loveliness of the popular person for the observer	
Daneshfar & Nabizadeh (2015), Forentnburi & Mac Goldric)(2011	familiarity of famous person with the observer	
Mansouri Moayed (2005), Aghapour (2016)	emotional music	
Khonsazadeh (2010), Hosseinzadeh et.al (2016), Aghapour (2016)	using various methods	
Siadat & Sakhi Jourashk (2014), Sohrabi Ranani (2016)	high coverage and being universal	
Navaei et.al (2016), Mosavi Esmaeilzadeh (2015), Hosseinzadeh et.al (2016), Mirzadeh & Mehrmanesh (2016), Rosemary et.al (2012), Mikit et.al (2012), Spenkar (2013)	media advertisement	
Navaei et.al (2015), Mosavi Esmaeilzadeh (2015), Hosseinzadeh et.al (2016), Ashouri (2016), Hassanzadeh & Danaei (2016), Mikit et.al (2012)	multimedia advertisement	
Navaei et.al (2015), Mirzadeh & Mehrmanesh (2016)	radio advertisement	
Navaei et.al (2015), Mosavi Esmaeilzdeh (2015), Ashouri (2016), Hassanzadeh & Danaei (2016)	cell phone advertisement	
Navaei et.al (2015), Hosseinzadeh et.al (2016), Mirzadeh & Mehrmanesh (2016)	creating TV advertisement	
Ahmadi (2013), Salimi Fard & Pourdehghan (2016), Hoang et.al (2013)	attraction of rewards	
siadat & Sakhi Jourashk (2014), Koumra et.al (2016)	being benevolent	

Darzaian (2016), Zabkouich & Louksten (2011), Abdolsalam et.al (2016)	coordination in speaking and behavior	
Ahmadi (2013), Salimi Fard & Pourdehghan(2016), Honari et.al (2015)	reality of messages	
Madhoushi (2013), Salimi Fard & Pourdehghan (2016), Houang et.al (2013)	graphic-designing	
Shahmohammadi (2011), Madhoushi (2013), Salimi Fard & Pourdehghan(2016)	psychological – geographical division	
Mirzadeh & Mehrmanesh (2016), Zabkouich & Lokesten (2011), Menroz et.al (2012), Choubhai et.al (2013), Mokan & Oltz (2014), Ayono et.al (2014)	advertisement on web	
Dehdashti Shahrokh (2009), Shahmohammadi (2011), Mirzadeh & Mehrmanesh (2016), Menroz et.al (2012), Choubhai et.al (2013), Mokan & Oltz (2014), Ayono et.al (2014), Henri et.al (2015), Yazghiro- Basteh (2015), Lin (2016)	internet & cell phone advertisement	
Shahmohammadi (2011), Sanei et.al (2016), Menroz et.al (2012), Chobai et.al (2013), Ayono et.al (2014), Henri et.al (2015)	selecting advertising media	
Shahmohammadi (2011), Sanei et.al (2016), Howang et.al (2013), Henri et.al (2015), Yazghirdo-Yasteh (2015), Martina Santa et.al (2017)	artistic aspect of advertisement	
Kashizadeh (2010), Hoshivar(2014), Pelpin et.al (2017)	gender of speaker in the advertisement	
Hosseinzadeh et.al (2016), oribeh (2016), Madra et.al (2017)	text repetition	
Shahmohammadi (2011), Hassanzadeh & Danaei (2016), Abri & Joumar (2014), Ayouno et.al (2014), Oribeh (2016), Pachman (2016), Bak (2016), Madra et.al (2017)	advertisement	
Samsam Shariat (2007), Sohrabi Ranani (2016), Daneshvar et.al (2016), Bak (2016), Kiji-Kali (2017)	factors relating to usage	environment

Fakhari et.al (2012), Mirzadeh & Mehrmanesh (2016), Howang et.al (2013), Bak (2016), kiji-Kali (2017)	factors relating to facilitation	
Dargi & Mohammadian (2010), Hamidizadeh (2013), Darzban (2016), Belman et.al (2012)	awareness of different status of the environment status	
Dargi & Mohammadian (2010), Mirzadeh & Mehrmanesh (2016)	awareness of access location of objective group to the product	
Dargi & Mohammadian (2010), Salimifard & Pourdehghan (2016)	awareness of norms	
Dargi & Mohammadian (2010), Hamidizadeh (2013), Darzban (2016), Belman et.al (2012), Koumra et.al (2016)	awareness of religions and religious values	
Dargi & Mohammadian (2010), Hassanzadeh & Danaei (2016)	awareness of method of evaluating services in relation to other services	
Hassazadeh & Danaei (2016), Radmiker (2015)	attention to security	
Samsam Shariat (2007), Mirzadeh & Mehrmanesh (2016), Belman et.al (2012)	economic beliefs	
Salimifard & Pourdehghan (2016), Pilay et.al (2012), Howang et.al (2013), Komra et.al (2016)	ethical beliefs	
Salimifard & Pourdehghan (2016)	Islamic teaching	
Samsam Shariat (2007), Daneshvar et.al (2016), Radmiker (2015)	strategic look to advertisement plan	
Dargi & Mohamadian (2010), Ansari & Rasei (2016), Abri & Joumar (2014)	market share	
Dargi & Mohamadian (2010), Darzain (2016)	obviating primary needs of audiences	

Salimi Fard & Pourdehghan (2016), Kim et.al (2012), Howang et.al (2013)	reference groups	
Sohrabi Ranin (2016), Fontini et.al (2009), Belman et.al (2012)	awareness of services	factors relating to services
Salimi Fard & Pourdehghan (2016), Abri & Joumar (2014), Radmiker (2015)	stating potential benefits of purchase ad services consumption	
Dargi & Mohamadian (2010), Hosseinzadeh et.al (2016), Fontini et.al (2009)	rate of services complexity	
Samsam Shariat (2007), Darzaian (2016), Fontini et.al (2009), Radmiker (2015)	display of expected experience	
Khashizadeh (2010), Darzain (2016)	rate of familiarity of consumer with services	
Ansari & Rasaei (2016), Belman et.al (2012)	market researches	
Hosseinzadeh et.al (2016), Mirzadeh & Mehrmanesh (2016)	record of exploiting services	
Dargi & Mohammadian (2010), Ansari & Rasaei (2016), Lousi (2012), Park (2015), Marion et.al (2017)	being unique	
Dehdashti Shahrokh (2009), Daneshvar et.al (2016), Radmiker (2015), Jankoski et.al (2016)	price, benefits	
Daneshvar et.al (2016), Fontini et.al (2009), Jankoski et.al (2016)	relation with customer	
Samsam Shariat (2007), Daneshvar et.al (201), Fontini et.al (2009), Jankoski (2016)	supporting customer	
Dehdashti Shahrokh (2009), Ansari & Rasaei (2016), Fontini et.al (2009), Radmiker (2015), Jankoski (2016)	feedback	

2- Research plan

The present study is of descriptive applied type which has been performed with survey method. The statistical population of this study includes all managers and experts of Eghtesad

Novin Bank. The statistical population of the present study contains all people who are exposed to advertisements of Eghtesad Novin Bank across the country. For selecting samples, Cochran formula is used for unlimited societies that based on this formula, the samples number were determined 400 persons for each province. The sample people were selected based on the following sampling method. At first, using stratified sampling method , the samples were randomly selected from five geographical regions of the country (north, south, east, west and center) of nine provinces (Mazandaran, Khorasan Razavi, Zanjan, Chahar Mahal Bakhtiari , Kerman, East Azarbaijan , Kermanshah, Ilam and Tehran).

3- Variables and research model

In this study, fuzzy Delphi technique has been used for identifying main indicators. For ranking the final criteria and sub standards, multi-criteria decision making models and AHP model have been used with fuzzy approach. For analysis of obtained data, Excel software and coding in the environment of this software have been used. The results obtained from data analysis have been provided in the present chapter.

3-1 Identifying criteria and sub standards of factors effective on advertisement effectiveness

At the first step, the primary indicators were identified and selected. In the first stage, by reviewing the literature and performed studies, a large number of indicators were extracted from the literature. For screening the indicators and identifying the final indicators, fuzzy Delphi approach has been used. The expert views have been collected in the field of rate of significance of main indicators. Fuzzy average and defuzzification output of rates relating to indicators are presented in table 2. The rate of defuzzification greater than 7 is acceptable and every indicator which has the score higher than 7 is confirmed (Wu& Fang, 2011).

Table 2- fuzzy average and fuzzy screening of indicators of factors effective on advertisement effectiveness

Result	Crisp	mean	U	M	L	
acceptance	7.15	(2, 7.97, 9)	9	7.97	2	Q01
acceptance	7.84	(6, 8.01, 9)	9	8.01	6	Q02

Result	Crisp	mean	U	M	L	
acceptance	7.92	(6, 8.13, 9)	9	8.13	6	Q03
acceptance	7.39	(4, 7.83, 9)	9	7.83	4	Q04
acceptance	7.70	(6, 7.8, 9)	9	7.80	6	Q05
rejection	6.68	(2, 7.28, 9)	9	7.28	2	Q06
acceptance	7.71	(6, 7.82, 9)	9	7.82	6	Q07
acceptance	7.66	(6, 7.75, 9)	9	7.75	6	Q08
acceptance	7.02	(2, 7.78, 9)	9	7.78	2	Q09
acceptance	7.27	(2, 8.15, 9)	9	8.15	2	Q10
acceptance	7.09	(2, 7.88, 9)	9	7.88	2	Q11
rejection	6.85	(2, 7.52, 9)	9	7.52	2	Q12
acceptance	7.43	(4, 7.9, 9)	9	7.90	4	Q13
rejection	6.71	(3, 7.07, 9)	9	7.07	3	Q14
acceptance	7.17	(2, 8.01, 9)	9	8.01	2	Q15
acceptance	7.13	(2, 7.95, 9)	9	7.95	2	Q16
rejection	6.85	(4, 7.02, 9)	9	7.02	4	Q17
acceptance	7.43	(4, 7.89, 9)	9	7.89	4	Q18
acceptance	7.70	(6, 7.8, 9)	9	7.80	6	Q19
rejection	6.80	(3, 7.21, 9)	9	7.21	3	Q20
acceptance	7.75	(6, 7.88, 9)	9	7.88	6	Q21
rejection	6.62	(3, 6.93, 9)	9	6.93	3	Q22
acceptance	7.61	(6, 7.66, 9)	9	7.66	6	Q23
acceptance	7.22	(2, 8.08, 9)	9	8.08	2	Q24
acceptance	7.04	(2, 7.81, 9)	9	7.81	2	Q25
acceptance	7.71	(6, 7.82, 9)	9	7.82	6	Q26
rejection	6.65	(3, 6.97, 9)	9	6.97	3	Q27
acceptance	7.98	(6, 8.22, 9)	9	8.22	6	Q28
acceptance	7.70	(6, 7.79, 9)	9	7.79	6	Q29
rejection	6.54	(3, 6.81, 9)	9	6.81	3	Q30
acceptance	7.19	(4, 7.53, 9)	9	7.53	4	Q31
acceptance	7.79	(5, 8.18, 9)	9	8.18	5	Q32

Result	Crisp	mean	U	M	L	
acceptance	7.53	(4, 8.05, 9)	9	8.05	4	Q33
acceptance	7.05	(2, 7.82, 9)	9	7.82	2	Q34
acceptance	7.43	(4, 7.89, 9)	9	7.89	4	Q35
rejection	6.92	(3, 7.38, 9)	9	7.38	3	Q36
acceptance	7.02	(2, 7.78, 9)	9	7.78	2	Q37
rejection	6.91	(3, 7.37, 9)	9	7.37	3	Q38
rejection	6.96	(2, 7.68, 9)	9	7.68	2	Q39
rejection	6.99	(2, 7.74, 9)	9	7.74	2	Q40
acceptance	7.42	(4, 7.88, 9)	9	7.88	4	Q41
acceptance	7.10	(2, 7.9, 9)	9	7.90	2	Q42
rejection	6.59	(2, 7.13, 9)	9	7.13	2	Q43
rejection	6.55	(2, 7.08, 9)	9	7.08	2	Q44
acceptance	7.77	(6, 7.9, 9)	9	7.90	6	Q45
rejection	6.91	(2, 7.62, 9)	9	7.62	2	Q46
acceptance	7.09	(2, 7.88, 9)	9	7.88	2	Q47
acceptance	7.68	(6, 7.78, 9)	9	7.78	6	Q48
acceptance	8.19	(6, 8.53, 9)	9	8.53	6	Q49
rejection	6.88	(2, 7.57, 9)	9	7.57	2	Q50
acceptance	7.75	(5, 8.12, 9)	9	8.12	5	Q51
acceptance	7.03	(2, 7.8, 9)	9	7.80	2	Q52
rejection	6.96	(4, 7.19, 9)	9	7.19	4	Q53
acceptance	7.04	(2, 7.8, 9)	9	7.80	2	Q54
rejection	6.96	(3, 7.45, 9)	9	7.45	3	Q55
acceptance	7.02	(2, 7.78, 9)	9	7.78	2	Q56
acceptance	7.92	(5, 8.38, 9)	9	8.38	5	Q57
acceptance	7.00	(2, 7.76, 9)	9	7.76	2	Q58
rejection	6.90	(1, 7.86, 9)	9	7.86	1	Q59
acceptance	7.10	(4, 7.4, 9)	9	7.40	4	Q60
acceptance	7.86	(6, 8.04, 9)	9	8.04	6	Q61
rejection	6.97	(2, 7.71, 9)	9	7.71	2	Q62

Result	Crisp	mean	U	M	L	
acceptance	7.02	(2, 7.79, 9)	9	7.79	2	Q63
acceptance	7.73	(6, 7.84, 9)	9	7.84	6	Q64
acceptance	7.01	(2, 7.76, 9)	9	7.76	2	Q65
rejection	6.65	(3, 6.98, 9)	9	6.98	3	Q66
acceptance	7.01	(2, 7.77, 9)	9	7.77	2	Q67
acceptance	7.90	(6, 8.11, 9)	9	8.11	6	Q68
acceptance	7.55	(5, 7.83, 9)	9	7.83	5	Q69
acceptance	8.02	(6, 8.28, 9)	9	8.28	6	Q70
acceptance	7.95	(6, 8.17, 9)	9	8.17	6	Q71
acceptance	7.03	(2, 7.79, 9)	9	7.79	2	Q72
acceptance	7.65	(5, 7.98, 9)	9	7.98	5	Q73
acceptance	7.00	(1, 8, 9)	9	8.00	1	Q74
acceptance	7.04	(2, 7.8, 9)	9	7.80	2	Q75
acceptance	7.03	(2, 7.8, 9)	9	7.80	2	Q76
acceptance	7.30	(4, 7.7, 9)	9	7.70	4	Q77
rejection	6.88	(2, 7.57, 9)	9	7.57	2	Q78
acceptance	7.63	(4, 8.19, 9)	9	8.19	4	Q79
acceptance	7.65	(5, 7.98, 9)	9	7.98	5	Q80
rejection	6.59	(3, 6.88, 9)	9	6.88	3	Q81
acceptance	7.09	(2, 7.88, 9)	9	7.88	2	Q82
rejection	6.83	(2, 7.5, 9)	9	7.50	2	Q83
acceptance	7.12	(2, 7.93, 9)	9	7.93	2	Q84

From the 84 indicators which were assessed, 24 indicators which have acquired the average score less than 7 are omitted. So, 60 indicators are finally confirmed (Adelazaret.al, 1995).

The remained indicators will be finally examined in the format of a 60-question questionnaire with analytical hierarchy process.

3-2 determining the priority of factors effective on effectiveness of advertisement using fuzzy AHP technique

For determining the priority of major criteria, fuzzy analytical hierarchy (FAHP) has been used. The analysis process is as follows.

1- Dual comparison of major criteria based on objective and determining the weight of major criteria

2- Dual comparison of sub standards of each criterion and determining sub standards of each cluster

3- Multiplication of the weight of sub standards in the weight of respective criterion and determining the final weight of sub standards

For dual comparison of elements, 9-degree Saaty scale has been used. 9-degree Saaty scale has been provided by Thomas Saaty , codifier of analytical hierarchy theory. Also, in this study fuzzy approach has been used for making the values quantitative. So, Saaty fuzzy range has been used.

Table 3- fuzzy range equal to 9-degree Saaty scale in AHP technique. Habibi et.al , 2014, 77

reversed fuzzy equivalent	fuzzy equivalent	the word term of comparison status of i to j
(1,1,1)	(1, 1, 1)	Preferred Equally
$\left(\frac{1}{3},\frac{1}{2},1\right)$	(1, 2, 3)	intermediate
$\left(\frac{1}{4},\frac{1}{3},\frac{1}{2}\right)$	(2, 3, 4)	Preferred moderately
$\left(\frac{1}{5},\frac{1}{4},\frac{1}{3}\right)$	(3, 4, 5)	intermediate
$\left(\frac{1}{6},\frac{1}{5},\frac{1}{4}\right)$	(4, 5, 6)	Preferred Strongly
$\left(\frac{1}{7},\frac{1}{6},\frac{1}{5}\right)$	(5, 6, 7)	intermediate
$\left(\frac{1}{8},\frac{1}{7},\frac{1}{6}\right)$	(6, 7, 8)	very strongly Preferred
$\left(\frac{1}{9},\frac{1}{8},\frac{1}{7}\right)$	(7, 8, 9)	intermediate
$\left(\frac{1}{9},\frac{1}{9},\frac{1}{9}\right)$	(9, 9, 9)	Extremely Preferred

(source in papersL Khorshid & Zabihi, Lee my H.I; Wen-Chin Chen, Ching-Jan, Chang, 2008, 101)

The pattern of criteria and sub standards has been drawn in the figure using AHP technique. Also, the criteria and sub standards of the study have been named with numeric index so that the research flow is easily detectable.

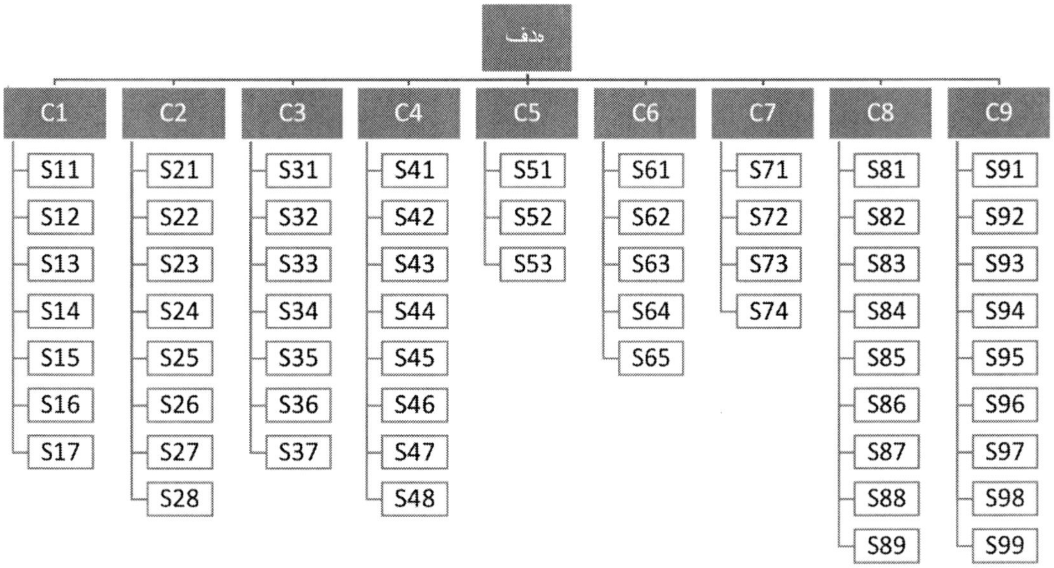

Figure 1- hierarchical model of criteria and sub standards of factor effective on advertisement effectiveness

1- Determining the priority of major criteria based on objective

In the first step, the major criteria based on the objective have been compared dually. Dual comparison is very simple and all elements of each cluster should be compared two by two. Therefore, if there is n element in a cluster, $\frac{n(n-1)}{2}$ comparisons will be performed. Since there are four criteria, so the number of performed comparisons is equal to:

$$\frac{n(n-1)}{2} = \frac{9(9-1)}{2} = 36$$

So, 36 dual comparisons have been performed from group perspective of experts. The experts view has been made quantitative using fuzzy scale. Collecting experts view has been collected with 9-degree Saaty range. So, 6 dual scales of four major four criteria have been performed based on the experts view. Using geometric average the experts views have been aggregated. Fuzzy aggregation will be calculated as follows:

Equation 1

$$F_{AGR} = (\prod(l), \prod(m), \prod(u))$$

After matrix formation of the obtained dual comparisons, the special vector has been calculated. At first, the fuzzy extension of each row is calculated. Every entry of dual comparison matrix $\tilde{X}$

Is displayed as $\tilde{x}_{ij}$.

The fuzzy extension of every row is also displayed with symbol $\tilde{S}_i$. Therefore, the fuzzy extension of each row is calculated as follows:

Equation 2

$$\tilde{S}_i = \sum_{j=1}^{n} x_{ij}$$

Then, the fuzzy extension of every row elements will be as follows:

Fuzzy extension of row 1 (1,1,1) ⊕ (1,27,1,47,1,75) ⊕ (0.47, 0.55, 0.66) ⊕ (1.52, 1.81, 2.14) ⊕ (0.93, 1.14, 1.39) ⊕ (1.26, 1.61, 2.04) ⊕(1.25, 1.51, 1.84) ⊕(0.78, 0.97, 1.24) ⊕ (0.59, 0.75, 0.96) ⊕ (9.08, 10.8, 13.02)

Fuzzy extension of row 2 (0.57, 0.68, 0.79)⊕(1, 1, 1)⊕(0.58, 0.7, 0.89)⊕(1.84, 2.28, 2.75)⊕(0.68, 0.82, 1.02)⊕(1.42, 1.81, 2.34)⊕(1.64, 2.04, 2.53)⊕(0.71, 0.88, 1.12)⊕(0.73, fuzzy extension of row 3 (1.52, 1.82, 2.11)⊕(1.13,)0.9, 1.1)⊕ = (9.17, 11.11, 13.55 1.42, 1.72)⊕(1, 1, 1)⊕(1.89, 2.34, 2.88)⊕(0.7, 0.86, 1.09)⊕(1.45, 1.78, 2.22)⊕(1.55, 1.88, 2.29)⊕(0.57, 0.71, 0.93)⊕(0.67, 0.8, 0.96)⊕ = (10.47, 12.61, 15.2)

Fuzzy extension of row 4 (0.47, 0.55, 0.66)⊕(0.36, 0.44, 0.54)⊕(0.35, 0.43, 0.53)⊕(1, 1, 1)⊕(0.97, 1.21, 1.51)⊕(0.96, 1.33, 1.8)⊕(2.14, 2.7, 3.37)⊕(0.94, 1.21, 1.56)⊕(1.3, 1.69, 2.16)⊕ = (8.49, 10.57, 13.14)

Fuzzy extension of row 5 (0.72, 0.88, 1.07) ⊕ (0.98, 1.22, 1.48)⊕(0.92, 1.17, 1.43)⊕(0.66, 0.82, 1.03)⊕(1, 1, 1)⊕(0.67, 0.83, 1.04)⊕(0.23, 0.29, 0.36)⊕(0.28, 0.32, 0.39)⊕(0.3, 0.38, 0.48)⊕ = (5.76, 6.91, 8.29)

Fuzzy extension of row 6 (0.49, 0.62, 0.79)⊕(0.43, 0.55, 0.7)⊕(0.45, 0.56, 0.69)⊕(0.55, 0.75, 1.04)⊕(0.96, 1.2, 1.48)⊕(1, 1, 1)⊕(0.34, 0.41, 0.51)⊕(0.27, 0.34, 0.44)⊕(0.24, 0.31, 0.41)⊕ = (4.74, 5.75, 7.08)

Fuzzy extension of row 7 (0.54, 0.66, 0.8)⊕(0.4, 0.49, 0.61)⊕(0.44, 0.53, 0.64)⊕(0.3, 0.37, 0.47)⊕(2.81, 3.5, 4.3)⊕(1.95, 2.43, 2.96)⊕(1, 1, 1)⊕(0.34, 0.42, 0.54)⊕(0.29, 0.36, 0.47)⊕ = (8.06, 9.76, 11.78)

Fuzzy extension of row 8 (0.8, 1.03, 1.28)⊕(0.89, 1.14, 1.41)⊕(1.08, 1.41, 1.76)⊕(0.64, 0.82, 1.06)⊕(2.55, 3.1, 3.63)⊕(2.25, 2.97, 3.72)⊕(1.86, 2.39, 2.94)⊕(1, 1, 1)⊕(0.58, 0.69, 0.83)⊕ = (11.66, 14.56, 17.64)

Fuzzy extension of row 9 (1.04, 1.33, 1.69)⊕(0.91, 1.11, 1.37)⊕(1.04, 1.25, 1.5)⊕(0.46, 0.59, 0.77)⊕(2.08, 2.63, 3.3)⊕(2.42, 3.23, 4.1)⊕(2.15, 2.74, 3.43)⊕(1.2, 1.46, 1.72)⊕(1, 1,)1)⊕ = (12.29, 15.35, 18.88

Then, the fuzzy sum of total priorities column elements is calculated

Equation 3

$$\sum \tilde{S}_i = \sum_{i=1}^{n}\sum_{j=1}^{n} x_{ij}$$

The sum of elements of the priorities column of major criteria will be as follows:

$$\sum_{i=1}^{4}\sum_{j=1}^{4} M_g^j = (79.72, 97.41, 118.59)$$

For normalization of each criterion priorities, that criterion total rates sum should be divided into sum of all priorities (column elements), since the rates are fuzzy, so fuzzy sum of every row is multiplied in summation reverse. The summation reverse should be calculated.

Equation 4

$$if\ \tilde{F} = (l, m, u)\ then\ \tilde{F}^{-1} = \left(\frac{1}{u}, \frac{1}{m}, \frac{1}{l}\right)$$

Then, based on the equation 4, we will have

$$\left(\sum_{i=1}^{n}\sum_{j=1}^{n} M_g^j\right)^{-1} = (0.01, 0.01, 0.01)$$

$$P_k = \sum_{i=1}^{n} M * \left(\sum_{i=1}^{n}\sum_{j=1}^{n} M_g^j\right)^{-1}.$$

So, the results obtained from normalization of obtained rates will be as follows:

$\widetilde{W}_{C1} =$ (0.08, 0.11, 0.16)

$\widetilde{W}_{C2} =$ (0.08, 0.11, 0.17)

$\widetilde{W}_{C3} =$ (0.09, 0.13, 0.19)

$\widetilde{W}_{C4} =$ (0.07, 0.11, 0.16)

$\widetilde{W}_{C5} =$ (0.05, 0.07, 0.1)

$\widetilde{W}_{C6} = (0.04, 0.06, 0.09)$

$\widetilde{W}_{C7} = (0.07, 0.1, 0.15)$

$\widetilde{W}_{C8} = (0.1, 0.15, 0.22)$

$\widetilde{W}_{C9} = (0.1, 0.16, 0.24)$

Each of the obtained rates is fuzzy and normalized weight relating to major criteria. There are various methods for defuzzifcation of the obtained rates. In this study, Bojadziev proposed method has been used for defuzzification.

Table 4- defuzzification and computation of final and normal weight of criteria of factors effective on advertisement effectiveness

Normal	Deffuzy	X3max	X2max	X1max	Crisp
0/111	0/117	0/113	0/115	0/117	C1
0/114	0/120	0/115	0/118	0/120	C2
0/103	0/137	0/133	0/135	0/137	C3
0/108	0/113	0/112	0/113	0/113	C4
0/070	0/073	0/072	0/073	0/073	C5
0/060	0/063	0/062	0/063	0/063	C6
0.101	0/107	0/103	0/105	0/107	C7
0.149	0/157	0/153	0/155	0/157	C8
0.158	0/167	0/163	0/165	0/167	C9

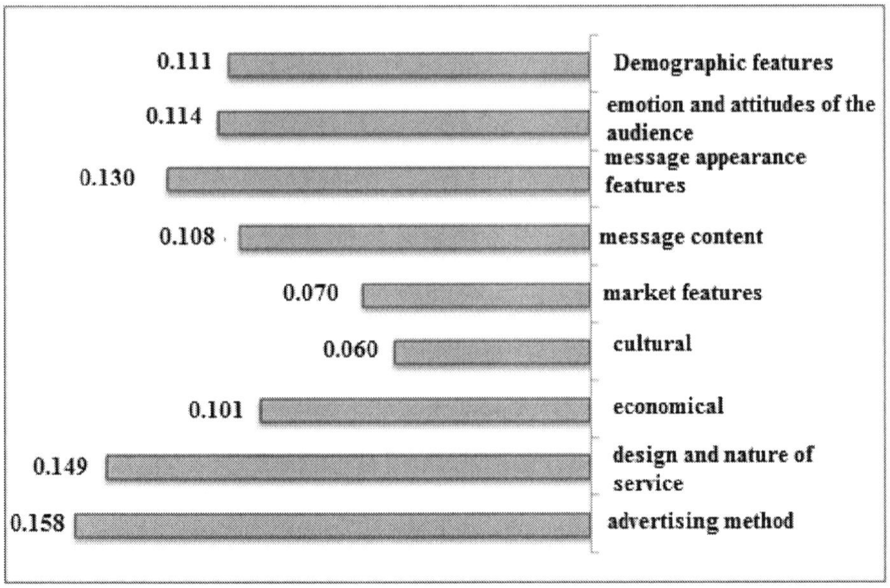

Figure 2- graphical display of significance of criteria of factors effective on advertisement effectiveness

Based on the obtained special vector:

- Demographic features with weight 0.111 have been located in rank five

- Individual feelings and attitude of the addressee with weight 0.114 has been located in rank four.

- The message surface features with weight 0.13 has been located in rank three.

- The message content with weight 0.108 has been located in rank six

- The market feature with weight 0.07 has been located in rank eight.

- Cultural with weight 0.06 has been located in rank nine.

- Economic with weight 0.10 has been located in rank seven.

- Plan and nature of services with weight 0.149 has been located in rank second.

- Advertisement methods with weight 0.158 has been located in rank first.

The inconsistency rate of performed comparisons has been obtained 0.030 which is smaller than 0.1 and so the performed comparisons is sufficed.

2- Comparison and determining priority of sub standards

At first, the experts' views have been collected in the dual comparison of sub standards. Then, the performed calculations for fuzzification of the experts views average have been provided for determining the priority of sub standards.

Priority of sub standards of demographic features

The audience job with weight 0.151 has been located in the third rank.

The audience income with weight 0.093 has been located in the seventh rank.

The audience religion with weight 0.16 has been located in the second rank.

Social class with weight 0.141 has been located in the fifth rank.

Awareness of gender differences with weight 0.173 has been located in the first rank.

The individual knowledge level with weight 0.135 has been located in the sixth rank.

Priority of sub standards of the audience individual feeling and attitude

Positive mentality with weight 0.124 has been located in the fifth rank.

Feeling of honesty in advertisements with weight 0.089 has been located in the eighth rank.

Threatening feeling with weight 0.14 has been located in the third rank.

Mental image with weight 0.102 has been located in the seventh rank.

Attitude towards brand with weight 0.121 has been located in the sixth rank.

Convincing audiences with weight 0.149 has been located in the first rank.

Attitude towards advertisement with weight 0.143 has been located in the second rank.

The individual perception of accuracy of advertisement with weight 0.133 has been located in the fourth rank.

Priority of sub standards of the message surface features

Then, based on the results of performed dual comparisons:

Beautiful and appropriate combination of colors with weight 0.143 has been located in the fourth rank.

Attention to reduction of the user waiting time with weight 0.159 has been located in the second rank.

Graphics- design with weight 0.131 has been located in the sixth rank.

Creativity in the message with weight 0.137 has been located in the fifth rank.

Emotional music with weight 0.099 has been located in the seventh rank.

Text repetition with weight 0.153 has been located in the third rank.

Access to required and sufficient information with weight 0.178 has been located in the first rank.

Determining the priority of sub standards of the message content

Benevolent with weight 0.093 has been located in the eighth rank.

Coordination in speech and behavior with weight 0.125 has been located in the fourth rank.

Reality of the messages with weight 0.119 has been located in the sixth rank.

Using women and children images with weight 0.131 has been located in the third rank.

Gender of speaker in the advertisement with weight 0.106 has been located in the seventh rank.

Using trustful and famous people with weight 0.16 has been located in the first rank.

Loveliness of famous people for the observer with weight 0.122 has been located in the fifth rank.

Familiarity of the famous person with the observer with weight 0.144 has been located in the second rank.

Priority of sub standards of the market features

Then, based on the results of the performed dual comparisons:

Factors relating to its application in the market with weight 0.326 have been located in the second rank.

Awareness of different aspects of status and environment with weight 0.283 has been located in the third rank.

Awareness of evaluation method of service receiver with weight 0.391 has been located in the first rank.

Priority of cultural sub standards

Then, based on the results of performed dual comparisons:

Awareness of norms with weight 0.199 has been located in the third rank.

Awareness of religion and religious values with weight 0.238 has been located in the first rank.

Ethical beliefs with weight 0.18 have been located in the fourth rank.

Obviating primary needs of the audiences with weight 0.164 has been located in the fifth rank.

Reference groups with weight 0.219 have been located in the second rank.

Determining the priority of economic sub standards

Then, based on the results of the performed dual comparison:

Economic beliefs with weight 0.261 have been located in the second rank.

Strategic look at an advertising program with weight 0.215 has been located in the fourth rank.

Market share with weight 0.274 has been located in the first rank.

Attention to economic security with weight 0.251 has been located in the third rank.

Priority of the plan and services nature sub standard

Awareness of services with weight 0.104 has been located in seventh rank.

Expressing potential benefits of purchase and services consumption with weight 0.117 has been located in the third rank.

Services complexity rate with weight 0.089 has been located in the ninth rank.

Uniqueness with weight 0.108 has been located in the sixth rank.

Rate of familiarity of the services consumer with weight 0.12 has been located in the second rank.

Price and allowances with weight 0.114 has been located in the fourth rank.

Relation with the customer with weight 0.111 has been located in the fifth rank.

Displaying the expected experience with weight 0.098 has been located in the eighth rank.

Supporting customers with weight 0.139 has been located in the first rank.

Priority of sub standards of the advertising methods

Beginning the advertisement with a challenge with weight 0.079 has been located in the ninth rank.

Beginning an advertisement with a statement and explanation with weight 0.107 has been located in the fifth rank.

Using stimuli with diversions more than usual and expected with weight 0.097 has been located in the seventh rank.

The advertisement message text being predicating with weight 0.126 has been located in the third rank.

Using diverse methods with weight 0.088 has been located in the eighth rank.

Media advertisement with weight 0.113 has been located in the fourth rank.

Dimensions of TV advertisement with weight 0.151 have been located in the first rank.

Psychological, geographical division with weight 0.107 has been located in the fifth rank.

Advertising media selection with weight 0.132 has been located in the second rank.

Determining the final priority of major criteria with FAHP technique

For determining final priority of major criteria using FAHP technique, the weights relating to major criteria (W1) and weight of indicators based on each criteria (W2) are needed. The results of comparison of the research sub criteria relating to them forms matrix W2. For determining the final priority of indicators with AHP technique, it is sufficient that the weight of indicators based on every criterion (W2) is multiplied in the major criteria (W1). Any of these matrices has been calculated in the previous steps. Based on the obtained output, it is obvious that the market share item, one of economic factors with weight 0. 277 is the most important indicator in the existing indicators. Item of awareness of evaluation method of service receiver in relation to other services with weight of 0.0273 has the second priority. Item of economic believes, one of economic factors, has the third priority.

So, the research model is as follows:

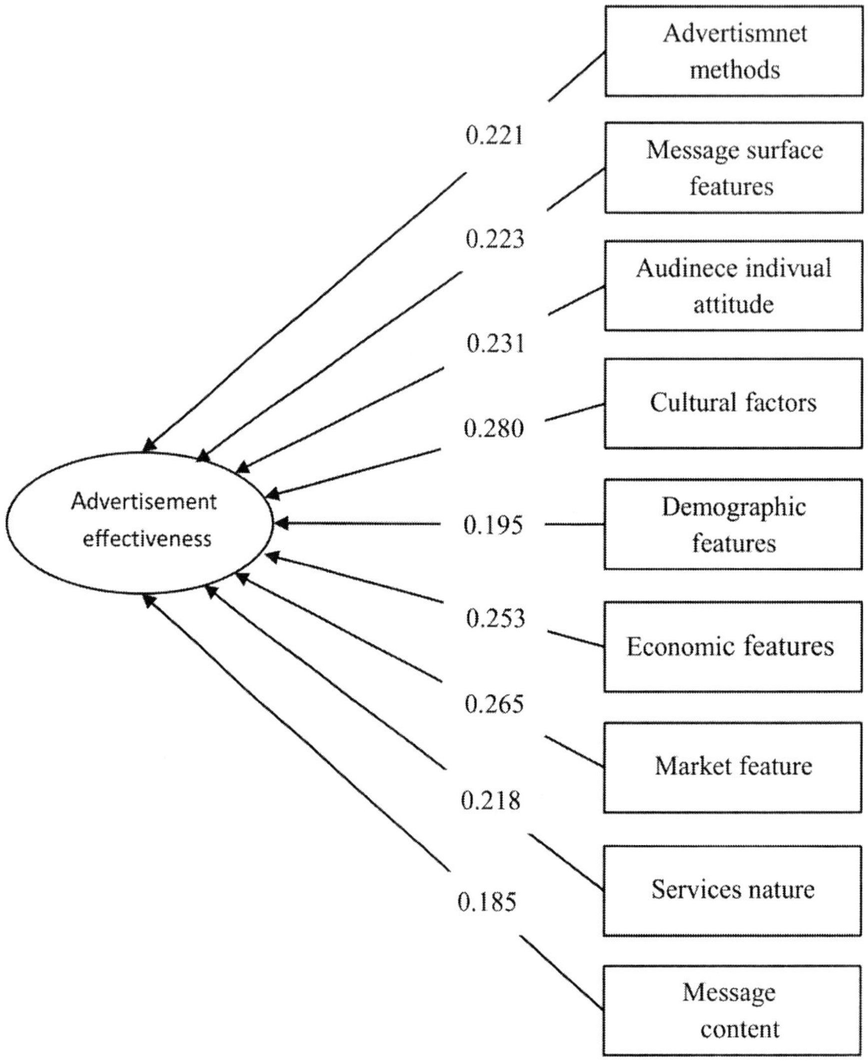

4- Conclusion:

In this study, identification and ranking of factors effective on designing the native model of advertisement in Eghtesad Novin bank has been addressed. In this first step, the primary indicators were identified and selected. In the first stage, by reviewing the literature and performed studies, a great number of indicators were extracted from the literature. For screening the indicators and identifying final indicators, fuzzy Delphi approach has been used. The experts' views have been collected in the field of significant rate of major indicators. From 84 indicators which were assessed, 24 indicators which have acquired score less than 7 are omitted. So, 60 indicators were finally confirmed. Then, regarding the performed calculations, the final weight of any of the model indicators have been calculated with fuzzy AHP. Based on the obtained output, it is obvious that the market share of economic factors with weight of 0.027 is the most important indicator in the existing indicators. The item of awareness of evaluating services in relation to other services is one of market features factors with weight 0.0273 which has second priority. The item of economic beliefs with weight 0.0264 is in the third priority.

Today, considering extension of competition in the international level and formation of world competition, all countries of the world may be successful in this arena to coordinate them with this trend and create more value for their customers by services and products they provide. In the today modern world, almost all people are exposed to advertisements and other communicational activities. Effectiveness of advertisement is one of important issues in the advertisement domain. In this regard, Mansouri Moyed (2005) showed that attention attraction, commercial advertisements of Maskan Bank are effective but in respect of other variables, that is, creating interest, stimulation of tendency and measurement to purchase, commercial advertisements of Maskan Bank are not effective. Samsam Shariat (2007) showed that TV channels and cartoon animations , attractive package, colors relating to the product and innovation, noticing safety feeling and associated symbols, informing of quality, background and price, dialects and common phrases, diversity in the message content, in TV, broadcasting from 18-23 and before beginning programs (in media) have been effective factors in effectiveness of commercial effectiveness from people viewpoint. Mosavi Esmaeilzadeh (2015) showed that oral advertisements (in 4 dimensions) and after that TV advertisement (just in 2 dimensions) are effective. Also, radio, internet and billboard advertisement are not effective. And also Friedman test was used for ranking advertisement tools. Salimi Fard & Pourdehghan (2016) showed that among factors effective on attitude to cell phone advertisement, components of general attitude towards advertisement, knowledge level and value have respectively the greatest significance and components of time of presentation and reference groups with the least significance in effecting the individuals attitude towards cell phone advertisement. Ansari & Rasaei (2016) showed that the creativity of advertisement message , market research, competition, market share, uniqueness and relation with customer are effective on the effectiveness of brand advertisement. Louis (2012) showed that advertising in the urban public spaces will have a great impact in the consumers' mentality. Since some advertisements are transitory and are often displayed in a short period of time and just in several spaces, but advertisement in public spaces exist everywhere and is similar to advertising through print and media, but since they have lower cost for production and distribution, have lower cost in this respect. Zabckowich & Loukston (2011) showed that in necktie advertisements use more of sensory requests and information signs. Findings showed that in western and European countries, people establish more contact due to globalization increase. They declared that content of printed advertisements, includes visual and text advertisements which are effective in establishing relation with many audiences. Also, ethnic identity as an important cultural variable increases attention to consumer purchase behavior. Besides that, they found out that the image of people, models, ceremonies and managers constitute an important part of images and background image plays an important role in effectiveness of visual factors in advertisement. Mac Ki- Nesbiat et.al (2011) showed that younger adults remind emotional messages specially negative ones better than others and older adults more prefer positive and logical messages to negative and emotional ones, but attitude towards advertisement among younger adults is not different. Also, they found that older adults have the most positive attitude towards advertisements.

Perzouti (2015) showed that advertising models are different in various ages and each age is consistent with a special model and using various and consistent models with people conditions may make the advertisement more effective. Abri & Joumar (2014) showed that advertising messages are effective on the consumer behavior and the advertiser may use various media for communicating consumer and media schedule is effective on the advertisement success. Attention to environmental factors where the customer lives are important factors in the advertisement success. Madra et.al (2017) showed that the advertisement text, text repetition impact the advertisement effectiveness. Maryion et.al (2017) showed that simultaneous use of several media like TV and internet may increase effectiveness of the advertisement and considering two variables of gender and multimedia create a better individual attitude towards the message effectiveness.

In ranking the identified factors, advertisement methods have assigned the first rank among factors. It is suggested that in designing advertisement and selecting advertising instruments, the dimensions of TV advertising, selecting median and predicative message text should be noticed. Using advertising instrument like internet advertisement, adverting on public vehicles like buses and metro due to great number of people who use them can increase stability of the advertisements in the customers mind.

In codifying and designing bank advertisements, the design and services nature should be noticed. It is proposed that supporting customer, familiarity of consumer with services are considered and during the advertisement process, it is required that the potential benefits of purchase and services consumption are expressed.

In designing the advertisement and content of bank advertising message, it is required to use famous and trustful people for introducing the services. Also, children as future customers and audiences should be noticed since they are future potentials. Total advertisements provided by the bank through various media should move in a unit line so that the audience can establish a logical relation between the observed advertisements in various media that is total advertising efforts lead to synergy of advertisement effectiveness. This issue won't be acquired unless through a media planning and efficient advertisement.

In advertisements, the message surface features should be noticed. It is required that in advertisement, access to required and enough information for the customers, attention to reduction of expectation time of user for observing the advertisement and suitable repetition of the text are considered. Providing large information in an advertisement which has a short time and providing information that require special mathematical analysis should be avoided.

References:

Hosseinabadi, Maryam, Mahmoud Nematian, Mohammad Besharati, investigating the relation between mixed elements of marketing and customers satisfaction with Tehran

Sarmayeh bank branches, management and entrepreneurship studies, fall 2016, 2nd period, No. 3 (page 162-169)

Dargi Parviz, 2008, subjects and topics of marketing management with Iran market attitude, Tehran, Rasa publication

Sarmad, Zohreh, Bazargan Abbas, Hejazi, Elaheh, 2014, research methods in behavioral sciences, Tehran, Agah. 27-13

Salimi Fard, Khodakaram, Pourdehghan, Adel, 2016, identifying and ranking factors effective on attitude towards cell phone advertisement using AHP, Brand management, period 3, no. 6, summer 2016, page 133-160

Seyed Akbar Manjgani, Vahideh Tabasi Lotfabadi, 2015, factors effective on effectiveness of advertisement on financial stoop of legal clients of Mashhad City with AHP approach, finance journal, fall 2015, No. 75 (page 121-146)

Sanayei, Ali, Majid Mohammad Shafiei, Maedeh Amini Valashani, 2016, effectiveness of bank advertisement in customer attraction, commercial management searches, spring and summer 2016, No. 15 (page 185-210)

Mahmoudian, Mahmoud, Rabiei, Ali, Jamili, Bita. 2011, evaluation of effectiveness of Parsian bank advertisement and identifying the most important factors in increasing its effectiveness in Tehran city, modern marketing studies, summer 2011, No. 2 (page 17-40).

Manjagani, Seyed Akbar, Tabasi Lotfabadi, Vahideh, 2015, factors effective on advertisement effectiveness on financial stoop of legal clinets of Mashhad city with AHP approach, finance journal, fall 2015, No. 75, 1-121

Venous, Davar, Safaeian, Mitra, (2005), marketing bank services in Iran and bank services features, Tehran, Negah Danesh, 4th edition

Angulo, N. (2016). 10 digital marketing trends to watch in 2016 and beyond. Retrieved from: http://www.marketingdive.com/news/10-digital-marketing-trends-to-watch-in-2016-./and-beyond/411457

Bell Raoul, Buchner, Axel.(2018). Positive Effects of Disruptive Advertising on Consumer Preferences, Journal of Interactive Marketing 41 (2018) 1–13

Corvi. Elisabetta.(2010).The effectiveness of advertising: a literature review, 10th Global Conference on Business & Economics, ISBN : 978-0-9830452-1-2

Cramphorn Michael F.,(2011), Forum Gender effects in advertising, International Journal of .Market Research Society Vol. 53 Issue 2

Cronin, J. J. & Fox , G. L. (2010). "The implications of third party customer complaining for .advertising efforts". Journal of advertising. 39(2): 21-23

Decrop, A. (2007). The influence of message format on the effectiveness of print advertisements for tourism destinations. International Journal of Advertising, 26(4), 505–.525

Denis .A. Shakhov and ,Anna A. Panasenko.(2012).Evaluating Effectiveness of Bank Advertising in the Internet: Theory and Practice, World Applied Sciences Journal 18 (Special Issue of Economics): 83-90, 2012 ISSN 1818-4952

Denis 1,2 A. Shakhov and 3Anna A. Panasenko.(2012).Evaluating Effectiveness of Bank Advertising in the Internet: Theory and Practice, World Applied Sciences Journal 18 (Special Issue of Economics): 83-90, 2012

Durdana Ozretic-Dosen Ines Zizak , (2015),"Measuring the quality of banking services targeting student population", EuroMed Journal of Business, Vol. 10 Iss 1 pp. 98 - 117

Eckler, P., and Bolls, P. (2011). Spreading the Virus: Emotional Tone of Viral Advertising and Its Effect on Forwarding Intention and Attitudes. Journal of Interactive Advertising, .11(2): 1-11

Eric W.T. Ngai Ka-leung Karen Moon S.S. Lam Eric S. K. Chin Spencer S.C. Tao , (2015),"Social media models, technologies, and applications", Industrial Management & Data Systems, Vol. 115 Iss 5 pp. 769 – 802

Fotini Patsioura Maro Vlachopoulou Vicky Manthou, (2009),"A new advertising effectiveness model for corporate advertising web sites", Benchmarking: An International Journal, Vol. 16 Iss 3 pp. 372 - 386

Frederic Crane, , G. (2010). «Marketing FO Entrepreneurs: ConceptandApplication for new .Venture. New York: SAGE Pablication, Inc

Frolova. Svetlana.(2014).THE ROLE OF ADVERTISING IN PROMOTING A PRODUCT, Thesis CENTRIA UNIVERSITY OF APPLIED SCIENCES ,Degree Programme in Industrial Management May 2014, p"1-61

Henry Boateng Abednego Feehi Okoe , (2015),"Consumers' attitude towards social media advertising and their behavioural response", Journal of Research in Interactive Marketing, Vol. 9 Iss 4 pp. 299 -312

Henry Boateng Abednego Feehi Okoe , (2015),"Consumers' attitude towards social media advertising and their behavioural response", Journal of Research in Interactive Marketing, Vol. 9 Iss 4 pp. 299 -312

Huang. Jinsong, Song Sub, , Liuning Zhouc, , , Xi Liu .(2013).Attitude Toward the Viral Ad: Expanding Traditional Advertising Models to Interactive Advertising, Journal of Interactive Marketing, Volume 27, Issue 1, February 2013, Pages 36–46

Idris, I., M.S.A. Yajid and A. Khatibi, 2009. Personal and Psychological Factors-Does it .Impact the Choice of Advertising Medium? Journal of Social Sciences, 5(2): 104-111

Ioannou Myria, Nikolaos Boukas, Evi Skoufari .(2014).Examining the role of advertising on the behaviour of co-operative bank consumers, Journal of Co-operative Organization and Management, Volume 2, Issue 1, June 2014, Pages 24-33

Jankowski. Jarosław, Przemysław Kazienko, Jarosław Wątróbski, Anna Lewandowska, Magdalena Zioło.(2016).Fuzzy multi-objective modeling of effectiveness and user experience in online advertising, Expert Systems with Applications, Volume 65, 15 December 2016, Pages 315-331

Kim, C., Park, S., Kwon, K., & Chang, W. (2012). An empirical test to measure the effectiveness of online advertising in online marketplaces using a hierarchical Bayes model. .Expert Systems with Applications (39), 117–128

Kim, ji-Hern., Hyun, yong j., (2010). " A model to investigate the influence of marketing – mix efforts and corporate image on brand equity in the IT software sector " . industrial . Marketing Management. No 40,2011, pp 424-438, Emerald Group Publishing Limited

Kim, S., & Jun, J. (2016). The impact of event advertising on attitudes and visit intentions. .Journal of Hospitality and Tourism Management, 29, 1-8

Kirby, J. and Marsden .(2006). Connected marketing : the viral, buzz and word of mouth .revolution, Oxford : Butterworth-Heinemann, 2006

*Madera. Quetzali, Oscar Castillo *, Mario García-Valdez, Alejandra Mancilla..(2017). A method based on Interactive Evolutionary Computation and fuzzy logic for increasing the effectiveness of advertising campaigns, Information Sciences 414 (2017) 175–186*

Marion Garaus and Udo Wagner , Anna-Maria B¨ack.(2017). The Effect of Media Multitasking on Advertising Message Effectiveness, Psychology & Marketing, Vol. 34(2):)138–156 (February 2017

Martín-Santana J.D., E. Reinares-Lara , P. Reinares-Lara .(2017).Influence of radio spokesperson gender and vocal pitch on advertising effectiveness: The role of listener gender, Spanish Journal of Marketing – ESIC, Volume 21, Issue 1, February 2017, Pages 63-71

McKay-Nesbitt, J., V. Manchanda, R., C. Smith, M., & A. Huhmann, B. (2011). Effects of age, need for cognition, and affective intensity on advertising effectiveness. Journal of Business Research (64), 12–17

Moch, Nils (2013). Competition in fragmented markets: New evidence from the German banking industry in the light of the subprime crisis. Journal of Banking &; Finance, Volume 37, Issue 8, Pp 2908–2919

Moorthy, S. and Hawkins, S. (2005). Advertising repetition and quality perception. Journal of .Business Research, 58, 354– 360

Mucan. Burcu, Özeltürkay. Eda Yaşa.(2014).Social Media Creates Competitive Advantages: How Turkish Banks Use This Power? A Content Analysis of Turkish Banks through Their Webpages, Procedia - Social and Behavioral Sciences, Volume 148, 25 August 2014, Pages 137-145

Muda, M.; Musa, R.; Niana-Mohamed, R. and Borhan, H. (2014). Celebrity Entrepreneur .Endorsement and Advertising Effectiveness. Social and Behavioral Sciences, 130, 11-20

Oladele. Kehinde Joseph, Ogunnaike, Akinbode Mosunmola, Aka Deborah .(2016). Effective Advertising: Tool For Achieving Client-Customer Relationships, Researchjournali's Journal of Media Studies Vol. 2 | No. 1 February | 2016

Olsen, G., Pracejus, J. W., & O'Guinn, T. (2012). Print advertising: White space. Journal of .Business Research (65), 855–860

Park, M. (2006). The compensatory effects of pictorial and verbal information for haptic information on consumer responses in non-store shopping environments. Unpublished Ph.D, .Ohio, Ohio state University

Park, T. (2008). Effective design of advertisement in handheld devices (Doctoral dissertation .)PURDUE UNIVERSITY

Park, T. (2008). Effectiveness Design of Advertisement in Handheld Devices. Dissertation for .doctoral degree, Purdue University, West Lafyette India

Pergelova, A., D. Prior and J. Rialp, 2010 Assessing Advertising Efficiency Does the Internet Play a Role? Journal of Advertising, 39(3): 39-54

Pilelienė Lina Viktorija Grigaliūnaitė , (2017)," The effect of female celebrity spokesperson in FMCG advertising: neuromarketing approach ", Journal of Consumer Marketing, Vol. 34 - .Iss 3 pp

Rosemary J. Averya, Matthew D. Eisenbergb, Kosali I. Simon .(2012).The impact of direct-to-consumer television and magazine advertising on antidepressant use, Journal of Health Economics, Volume 31, Issue 5, September 2012, Pages 705–718

Selviaridis. Kostas, Martin Spring b, Luis Araujo b .(2013). Provider involvement in business service definition: A typology, Industrial Marketing Management 42 (2013) 1398–1410

Strebinger .Andreas, Xiaoling Guo, Ferdinand Klauser, Peter Grant-Hayd .(2018).Is Multi-Ethnic Advertising a globally viable strategy for a Western luxury car brand? A mixed-method cross-cultural study, Journal of Business Research 82 (2018) 409–416

Suvittawat, A., (2015). "Commercial Bank Service Quality: Chonburi Province, Thailand". .International Journal of Management Sciences and Business Research. 4(2): 2226-8235

Tafani. Eric, Elyette Rouxa, Rainer Greifeneder .(2018).In the mood for action: When negative program-induced mood improves the behavioral effectiveness of TV commercials, Journal of Business Research 84 (2018) 125–140

Te'eni-Harari, T. (2014). "Clarifying the relationship between involvement variables and .advertising effectiveness among young people". Journal of Consum Policy. 37: 183–203

Thomas, S. and Kohli, S.C. (2011). Can Brand Image Move Upwards after Sideways? A .Strategic Approach to Brand Placements. Business Horizons, 54 (1): 41-49

Yahyaoui . Hamdi, Own .Hala S..(2018). Unsupervised clustering of service performance behaviors, Information Sciences 422 (2018) 558–571

Yuksel, A., Yuksel, F. & Bilim, Y. (2010) "Destination Attachment: Effects on Customer Satisfaction and Cognitive, Affective and Cognitive Loyalty", Journal Of Tourism Management, Vol.31, pp: 274-284.

STUDY OF NATURAL ANTIOXIDANT INHIBITORS FOR CORROSION PROTECTION OF MILD STEEL IN ACIDIC MEDIUM

Ms. Ibtihal Hassan AL-Bahrani, National University of Science and Technology, Sultanate of Oman.

ABSTRACT

Corrosion is a natural phenomena and it is a very big loss for structure as well as business; so corrosion minimization is one of the biggest areas of research. In this project; corrosion inhibition efficiency of berries juice extract was tested in mild steel sample against the sulfuric acid medium which was prepared from four types of berries; blackberries, blueberries, cranberries, and raspberries.

The experiment was designed by using Design Expert V 11 where immersed time, acid and berries juice extract concentrations were set as factors and a limited range was selected. The corrosion rate was set as a response. 20 runs were under taken and the experiments were done in the lab with maximum care.

The corrosion rate of 1020.374 mpy minimized to 597.659 mpy in presence of 5ml of extract and the efficiency of berries juice extract reached to 41.427%. The RSM gave 232.022 mpy corrosion rate as a minimum value and the optimum equation given by RSM was used for Genetic Algorithm analysis which gave 188.866 mpy corrosion rate at the optimizing conditions.

The surface and chemical characterization of the mild steel sample in presence and absence of berries juice extract were analyzed along with the relations between temperature, immersed time, acid and berries juice extract concentrations with response. The mixing tank was designed, calculations and drawing were presented.

KEYWORDS

Corrosion, Mild steel, Berries, Inhibitor, Weight loss, Sulfuric acid, RSM, SEM.

INTRODUCTION

Mild steel is used in most of our daily life applications as well as in industrial applications. In other hands, mild steel has low corrosion resistance which corrodes easily once it reacts with the surrounding environment; corrosion is a phenomenon where the mild steel reacts with the deteriorates and surrounding environment. Corrosion of mild steel is a continuous and constant problem where it severe financial and life loses as it leads to loss in the product, efficiency of equipment and failures in some cases (Roberge,2012).

The corrosion phenomenon controlled by adding a small amount of component to the corrosive environment which is a corrosion inhibitor. There are organic and inorganic inhibitors. The inorganic inhibitors are used in limit amount as they are toxic such as nitrites, molybate, nitrate, silicate, chromate, and phosphate (Cafferty,2010). Now a day, the corrosion engineers looking for cheap, renewable, friendly, readily available and ecologically acceptable inhibitors which are organic inhibitors which comes from the plants such as antioxidant components consisting of tannins, saponins, flavonoids, alkaloids and amino acids. Organic inhibitors consist of some heteroatoms such as oxygen, nitrogen, multi-bound, and sulphur atoms with the conjugated system which allows the absorption in the metal surface. Those atoms attract the absorption of some components on the metal surface and that done by the interaction of metals and by the electron of the conjugated system or by forming a bond with the free electron of heteroatom which forms a layer in the metal surface, as a result, it decreases the corrosion rate (Roberge,2012).

Berries are small pulpy fruits which have up to 9.2 mmol of antioxidant per 100g of berries. Berries can be used as a corrosion inhibitor as they are rich in antioxidant. The studies show that the berries juice extract of blueberries, cranberries, raspberries, and blackberries rich with antioxidant where the antioxidant capacity of them reach to 10 mg/ 100 g of blueberries (Michalska&Lysiak,2015), 58 mg/100g of cranberries (Blumberg,J.,et al,2013), 26.2mg/ 100 g of raspberries (Bobinaite,R.,Viskelis&Venskutonis,2015), and 21 mg/100 g of blackberries (Kaume,L.,et al,2012).

OBJECTIVES

To prepare anticorrosive extract (berries), to study corrosion in sample in presences and absences of the inhibitors, to determine the optimum conditions to inhibit corrosion and to design mixing tank.

METHADOLOGY, DESIGN AND EXPERIMENTATION

PREPARATION OF MILD STEEL SAMPLES

The mild steel sheets were cut into 2.5*3 cm^2 by using steel cutter and cleaned by surface cleaning machine to remove any corroded products in the metal surface then 4mm hole done by using drilling machine. Bunches for all metal samples (1-20) were made to differentiate between them. The drilled samples were washed by distilled water then rinsed with benzene and acetone respectively and dried with a hot air dryer. They were weighted using sensitive balance and kept in the desiccator until the time of usage.

PREPARATION OF CORROSIVE MEDIUM

The corrosive medium of different concentrations of H_2SO_4 prepared by using 36.8 N of the sulfuric acid.

PREPARATION OF BERRIES JUICE EXTRACT

137g, 209g, 282g, and 200g of raspberries, cranberries, blueberries, and blackberries respectively have been used to produce 309.5 ml of berries juice extract. The different types of berries washed by cold water, crashed in the electrical crasher and compressed mechanically by hand then filtered by cloth filter. The filtrate filtered again by using filter paper. The berries juice extract kept in the freezer at 0°C till the time of usage.

The efficiency of berries juice extract has been tested by running three experiments. The cleaned and weighted mild steel samples immersed in 0ml(blank), 2ml and 5ml of berries juice extract respectively with 50 ml of 1M H_2SO_4. After 2 hours; the samples have been taken out and cleaned by distilled water, benzene and acetone then dried by air dryer and weighted again. The corrosion rate and inhibitor concentration have been calculated by using equations (1) and (2) respectively.

$$Cr(mpy) = 22300\times\frac{\Delta W}{\rho\times A\times t} \quad \textbf{equation(1)}$$

$$\mathrm{IE}(\%) = \frac{\text{initalcorrosionrate-finalcorrosionrate}}{\text{initalcorrosionrate}}\times 100 \quad \textbf{equation(2)}$$

DESIGN THE EXPERIMENT CONDITIONS

The experiments conditions have been designed by using Design Expert software-version 11 and following Response Surface Module(RSM); the range of the parameters have been given as inputs to the software which are inhibitor concentration (5-15 ml), immersed time (180-240 min) and acid concentration (1-2 M) and the response set as corrosion rate in mpy. Table(1) shows the experiment conditions given by the software.

Table(1): Designed runs given by RSM.

Run.No	Factor(A): inhibitor, ml	Factor(B): Time, min	Factor(C): Acid concentration, M
1	10	180	1
2	15	240	2
3	5	120	1
4	5	240	1
5	10	180	1.5
6	15	180	2
7	10	120	1.5
8	5	120	1.5
9	10	180	2
10	10	180	1.5
11	10	180	1.5
12	15	240	1
13	15	120	1
14	5	240	2
15	10	180	1.5
16	15	180	1.5
17	5	120	2
18	10	180	1.5
19	10	180	1.5
20	10	240	1.5

APPLYING THE DESIGNED RUNS IN THE LABORATORY

The weighted mild steel sample has been hanged by using a plastic thread in a 50 ml of sulfuric acid in presence of berries juice extract of known concentration and contact time as per the software; the batch has been kept in the shaker. After the contact time the mild steel sample cleaned by using 120 grade emery paper to remove any corroded products, washed by distilled water, rinsed by benzene and acetone respectively and dried by hot air dryer, weighted by using sensitive balance. The corrosion rate in mpy was calculated by using equation(1) where the weight losses results have been used.

STUDYING THE IMPACT OF TEMPERATURE IN CORROSION RATE

To study the impact of temperature in the corrosion rate; mild steel samples weighted and hanged for 2 hours in 50 ml of 1M sulfuric acid with 10ml of berries extract at different temperatures(30, 40, 50 and 60 °C) as shown in Table(2); each batch has been kept in the water bath shaker with heater to provide perfect contact between the solution and mild steel sample and to maintain the required temperature. Corrosion rate has been calculated by equation(1).

Table(2): Experiment conditions of temperature into corrosion rate.

SI.NO	Temperature (°C)	Inhibitor concentration (ml)	Acid concentration (M)	immersed time, (hours)
1	30	10	1	2
2	40	10	1	2
3	50	10	1	2
4	60	10	1	2

SURFACE AND CHEMICAL CHARACTERIZATION ANALYSIS

The mild steel surface of blank, absence and presence of berries juice extract samples have been studied by using Scanning Electron Microscopy (SEM) and Energy Dispersive Spectroscopy (EDS).

OPTIMIZATION STUDY BY GENETIC ALGORITHM

Matlab has been used for optimization where the genetic algorithm method has been followed. The actual equation form Design expert(RSM) has been used where "Optimtool" followed. The same limit of the conditions from RSM has been selected. The same previous procedure has been followed in the lab with maximum care by using those optimum conditions which given by the software, 1.001M of H_2SO_4 with 15 ml of berries juice extract immersed for 237.208 minutes.

DESIGN OF MIXING TANK

Design of mixing tank is a part of the project where the industrial scale mixing tank has been designed, the tank will be formed from high resistance material such as S.S to be able to resist the corrosive medium of sulfuric acid which will be mixed with berries juice extract. It is a container which used to mix various material (heterogeneous materials) together to form homogenous materials even the different viscosity materials can be mixed easily which provides continuous process, good control, easy cleaning, and simplest construction. It can be made from S.S material which provides a smooth surface and it can be easily cleaned (Petropedia,2018).

Parameters to be considered to design mixing tank are length, torque, weight, stiffness and hydraulic loading.(Dynamixianc,2015). The important factors in design mixing tank are tank type, tank volume, specific gravity, process and environment.(Ibryamoval,Y.et al,2009)

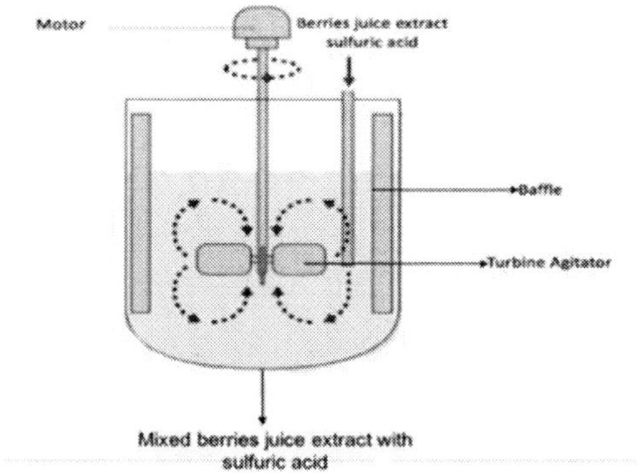

Figure(1): Mixing Tank(*Mixing Tank,*2018)

From the graph in Figure(2) Turbine agitator has been selected as it is suitable with the viscosity of the fluid and the tank volume. (Sinnot&Towler,2009)

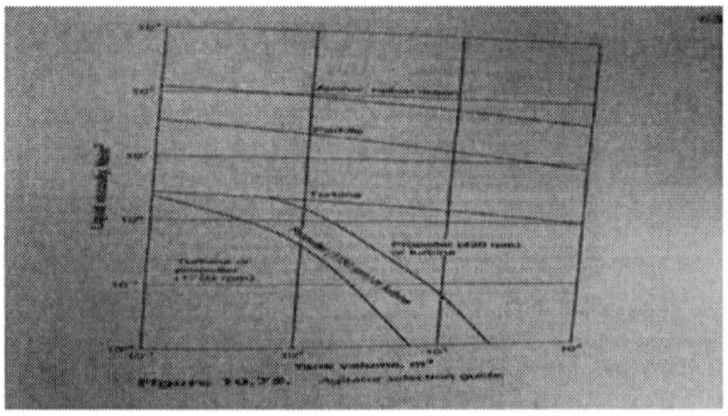

Figure(2): Graph of liquid viscosity Vs tank volume. (Sinnot&Towler,2009)

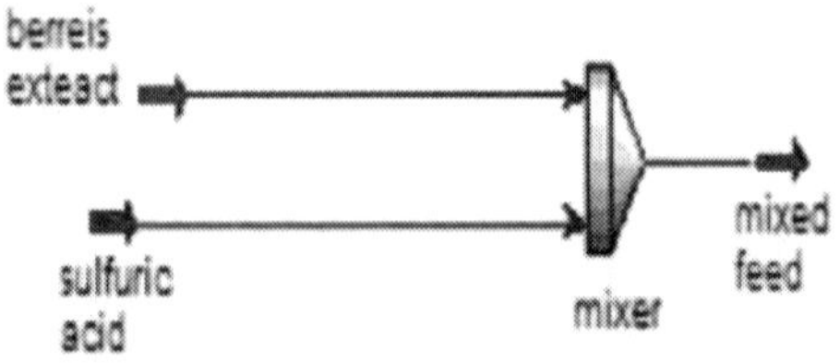

Figure(3): Result of Mixing Tank by Aspen Hysys.

Table(3): Calculated parameters for designing mixing tank

Tank diameter	m	3.5
Tank Height	m	1.81
Across sectional area	m^2	9.62
Volume of the tank	m^3	17.5
Impeller diameter	m	1.16
Length of paddle	m	0.29
Width of paddle	m	0.23
power	watt	204085
Reynolds Number	-	136486.6391

RESULTS AND ANALYSIS

EFFECT OF BERRIES JUICE EXTRACT CONCENTRATION INTO CORROSION RATE AND INHIBITOR EFFICIENCY

As shown in Table(4) the corrosion rate in the absence of berries juice extract inhibitor reach to the maximum (1020.374mpy); the corrosion rate has been minimized to 898.117mpy and 597.659mpy in presence of 2 and 5 ml of berries juice extract inhibitor respectively which means that berries juice extract inhibits the corrosion as the concentration of the berries juice extract increases. As well as the inhibitor efficiency reaches to 41.427% which obtained at 5ml of berries juice extract inhibitor in 1M of sulfuric acid immersed for 2 hours. The adsorption of the organic compounds such as ascorbic acid that present in the berries juice extract is blocking the active sites on the mild steel samples as it is forming a protective film on the metal surface as what kumar.A,2012 mention in his studies so the corrosion rate keeps decreasing as the amount of the berries juice extract is increasing. (Kumar.A,2012) Those compounds have heteroatoms such as oxygen which is a center of adsorption. In addition; It forms a coordinate bond as it has the ability to transfer the electrons to the vacant d-orbital of mild steel from the heteroatoms. (Ostovari, L. et al,2009)

Table(4): Experimental results of testing the efficiency of berries juice extract.

SI.NO	Inhibitor concentration (ml)	Corrosion rate (mpy)	Inhibitor efficiency (%)
1	0	1020.374	-
2	2	898.117	11.976
3	5	597.659	41.427

EFFECT OF IMMERSED TIME INTO CORROSION RATE AND INHIBITOR EFFICIENCY

The maximum corrosion rate reaches to 765.005mpy which obtained at 5ml of berries juice extract in the presence of 50 ml of 2 M of H_2SO_4 at 2 hours. Since the minimum corrosion rate of 232.022mpy obtained in presence of 5ml of berries juice extract with 50 ml of 1 M of H_2SO_4 immersed for 4 hours. The effect of immersed time into corrosion rate can be observed at low acid concentration and in the presence of berries juice extract, the corrosion rate is minimized at high immersed time. The decrease in the corrosion rate with the increase in immersed time is related to the adsorption of the corrosion product ($FeSO_4$) as shown in equation(3); According to Xianghong,L. et al ; the corrosion product of $FeSO_4$ becomes thicker with the immersed time, so it form a protective film (barrier) in the surface of mild steel (Xianghong, L. et al,2009).

$$Fe_{(s)}+H_2SO_{4(aq)} \rightarrow FeSO_{4(aq)}+H_{2(g)}$$, **equation(3)**

EFFECT OF ACID CONCENTRATION INTO CORROSION RATE

The effect of acid concentration into corrosion rate has been studied by using 1, 1.5 and 2 M of H_2SO_4 at different inhibitor concentrations and immersed time where the corrosion rate increase with the acid concentration at low inhibitor concentration and immersed time which reach to 765.005mpy at 2M H_2SO_4, 5ml of berries juice extract which immersed for 120 minutes as a maximum corrosion rate as shown in Table (3). In presence of the corrosive medium of H_2SO_4 with less amount of berries juice extract; the mild steel surface was exposed to H_2SO_4 medium where the corrosion reaction started as shown in equation(3) and $FeSO_4$ was formed with H_2 bubbles.

EFFECT OF TEMPERATURE INTO CORROSION RATE

From Table(5) can be noted that the rate of corrosion increases with temperature where it reaches to maximum 2407.994mpy at maximum temperature of 60°C. According to Ambrish, S. et al, at high temperature the berries juice extract molecules were partially absorbed in the mild steel surface so it was exposed to the corrosive medium of H_2SO_4, as a result, increasing the corrosion rate (Ambrish,S. et al,2012).

Table(5): Results of studying the effect of temperature into corrosion rate

SI.NO	Temperature,° C	Corrosion rate, mpy
1	30	701.507
2	40	1666.404
3	50	2008.861
4	60	2407.994

ANALYSIS BY RSM

Once the experimental results and corrosion rate calculated the RSM analysis has been done where the corrosion rate has been given to the software as a response. The response must be modified where the results fit with linear. The fit summary shows that the model F-value and P-value of 12.66 and 0.0001 respectively imply that the model is significant where A, B, C, and B^2 are the significant terms in this model only. 1.48 is a lack of fit P-value which indicates that the fit is not significant as there are pure errors.

0.5562 of the predicted R^2 is a reasonable agreement with 0.7105 of the adjusted R^2. The signal to noise ratio of 13.7252 of Adeq precision that the ratio grater than 4 is desirable. (Mycourses,2016)

The process variables are studied using ANOVA. The estimate coefficients in term of coded factors are A-Inhibitor, B-time, C-Acid and B^2. It is creating an accompanying condition for the sample with 1.0000 correlation coefficient. The coded and actual equations given by software are shown in equation(4) and (5) respectively.

- Corrosion rate(mpy)= 370.52-40.52(A)-126.97(B)+ 80.85(C) +66.58$(B)^2$ **equation(4)**
- Corrosion rate(mpy)= 1189.15487-8.10479(Inhibitor) -8.77421(Time) +161.69302 (Acid Conc)+0.018495 $(Time)^2$ **equation(5)**

The various relations can be studied from one factor graph; it can be noted that the corrosion rate increases with acid concentration and decreases with the increase in the amount of berries juice extract and immersed time as can be seen from the plots in Figure(3):Plot pf corrsoion rate Vs inhibitor concntration, Figure(4): plot of corrosion rate Vs acid concentration and Figure(5):Plot of corrosion rate Vs immersed time respectively.

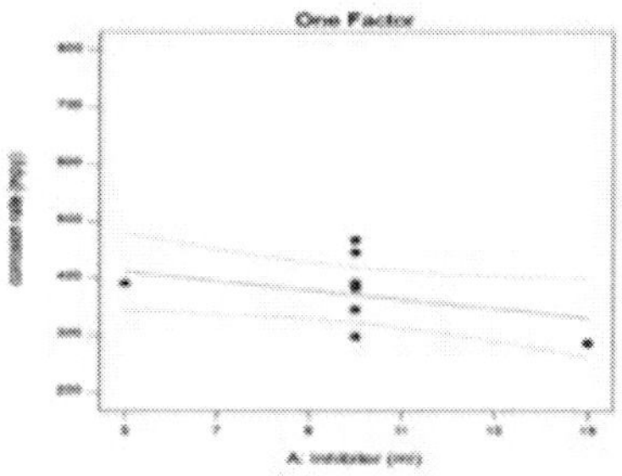

Figure(3)

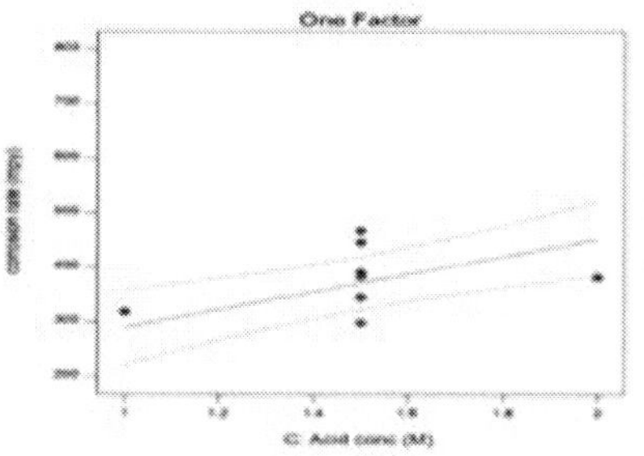

Figure(4)

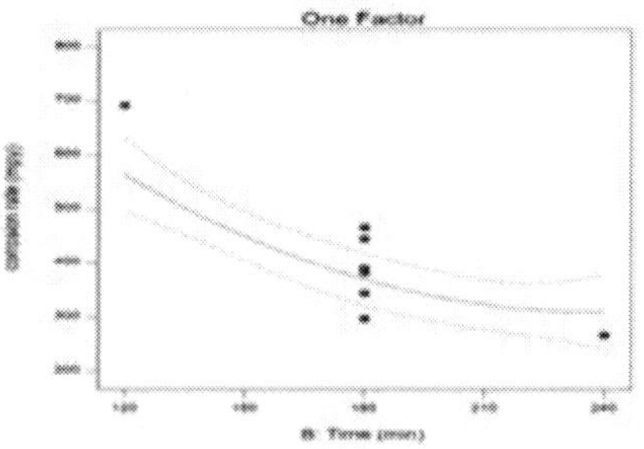

Figure(5)

The effects of two factors into the corrosion rate can be studied from the 3D plots as shown in Figure(6): 3D plot of corrosion rate Vs inhibitor concntration and acid concntration and Figure(7): 3D plot of corrosion rate Vs inhbitor concntration and immersed time.

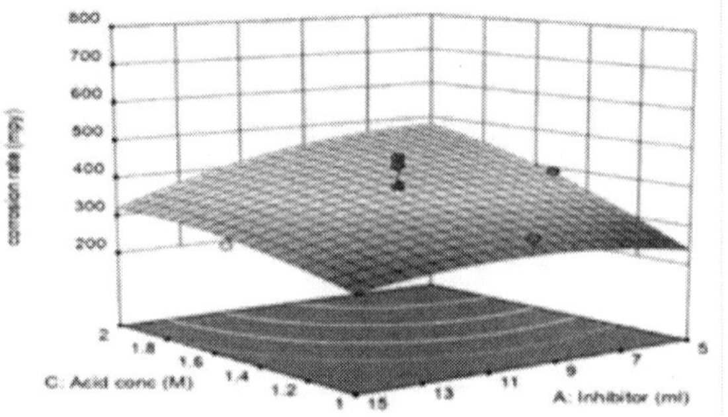

Figure(6)

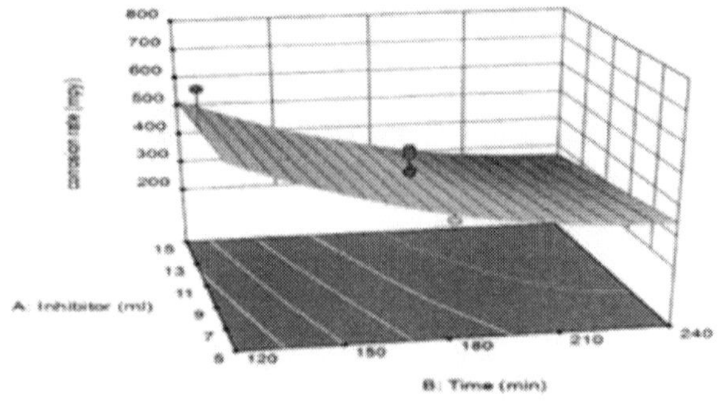

Figure(7)

The optimum conditions given by the RSM of 1M of H_2SO_4 with 5ml of berries juice extract inhibitor runs for 240 minutes gives 213.402mpy as a minimum corrosion rate but the experimental results at the same conditions gives 232.022mpy which means that they are too close as the suggested solution considered as an excellent optimum conditions.

SEM ANALYSIS

The SEM has been done for the blank mild steel sample for facilitating comparison as shown in Figure(8) which shows smooth mild steel surface compare to others. Rough surface in absence of berries juice extract is shown in Figure(9); where corrosion starting with the formation of H_2 bubbles as per the equation(3) which means that the surface of mild steel is exposed to the H_2SO_4 medium. In presence of 15 ml of berries juice extract, the surface of mild steel is protected by the adsorption of the organic compounds that are present in the berries juice extract where fewer cracks and pits can be noted and smooth surface is shown in Figure (10), which means that the mild steel surface is protected. (Raja, P. et al, 2008)

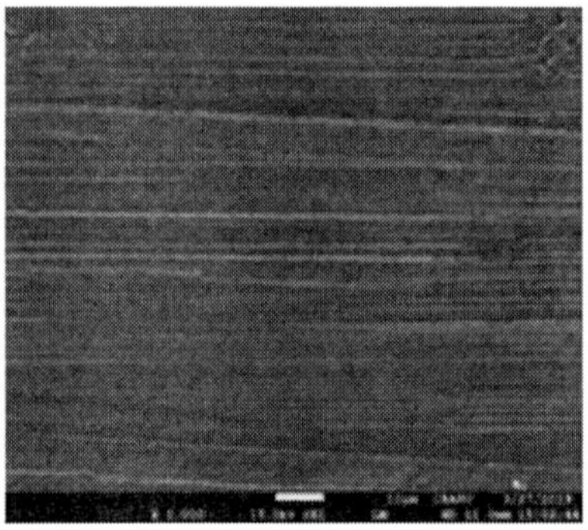

Figure(8)

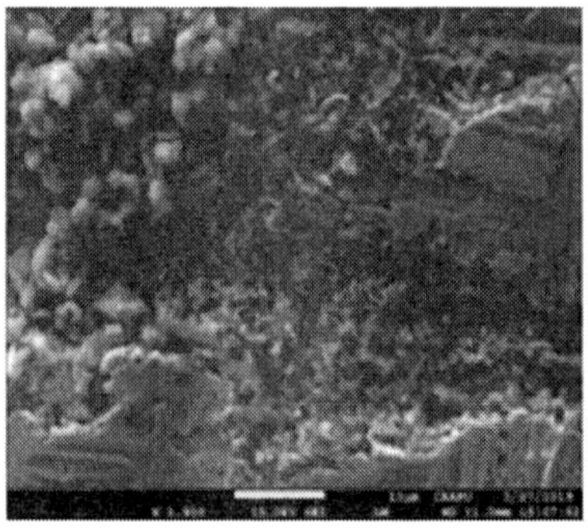

Figure(9)

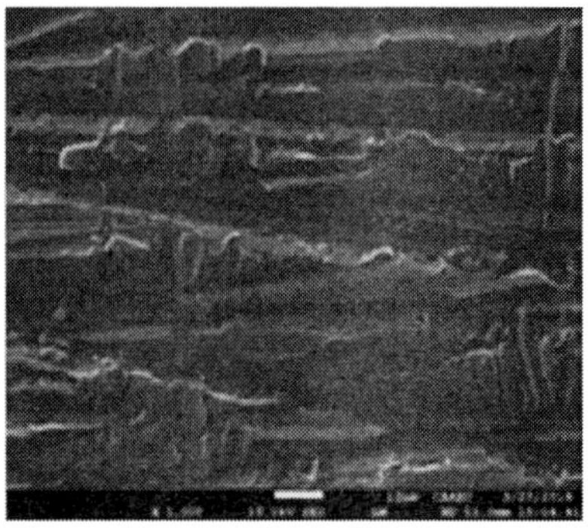

Figure(10)

EDS ANALYSIS

EDS has been used to study the chemical characterization or elemental analysis; Figure (11) shows the blank mild steel sample before being immersed in 1M H_2SO_4 in absence of berries juice extract where high Fe peak is shown, which indicates the presence of high amount of iron which reaches to 88 Wt% in the mild steel surface. Figure (12) shows the mild steel surface that has been immersed for 2 hours in 50 ml of 1M H_2SO_4 in absence of berries juice extract; the mild steel surface contains low amount of Fe (87.3 Wt%) compared to the blank mild steel sample. In Figure (13) the mild steel sample immersed for 2 hours in 50 ml of 1M H_2SO_4 in the presence of 5ml of berries juice extract, it shows high spark of Fe where the mild steel surface contains lower amount of Fe (84.8 Wt%) compared to Figure(11) and (12) which indicate lower iron dissolution rate in 1M H_2SO_4 medium containing berries juice extract (Alaneme, K. et al,2016). A high amount of oxygen and carbon are observed in presence of berries juice extract compared to the absence of berries juice extract. Si is shown in the absence and presence of berries juice extract which means that's it comes out to the surface during the reaction with sulfuric acid.

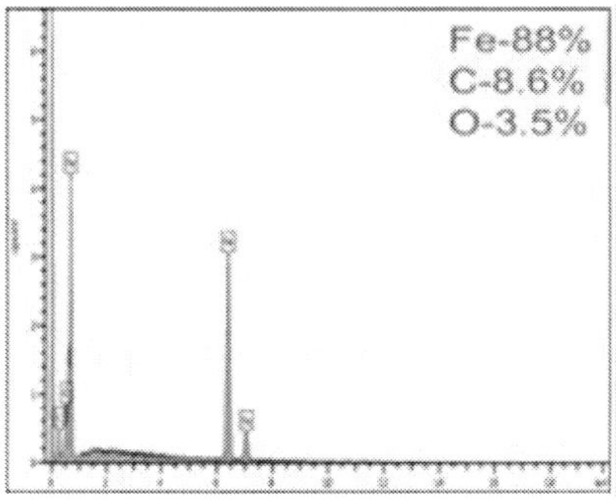

Figure(11)

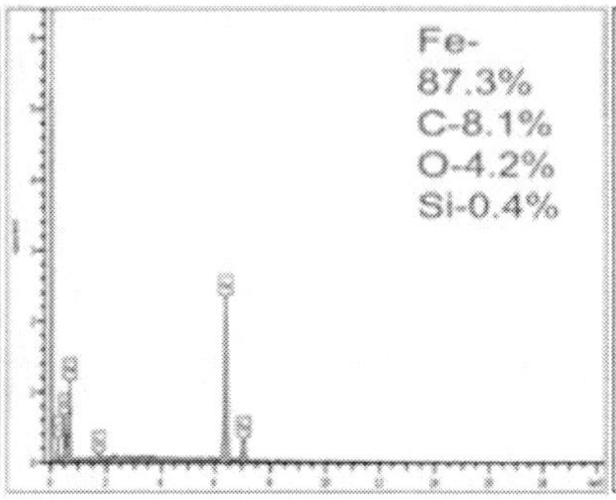

Figure(12)

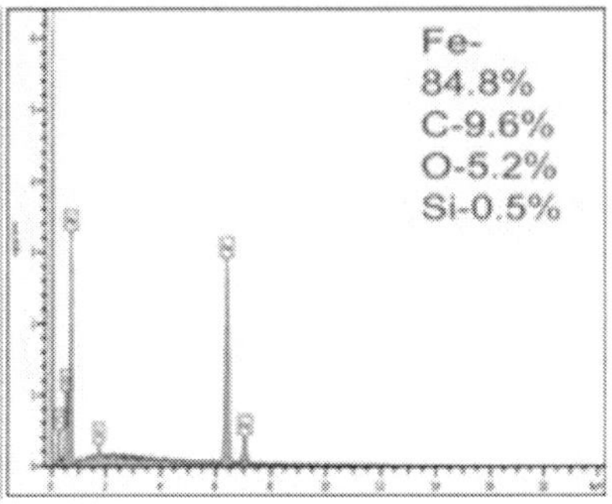

Figure(13)

GENETIC ALGORITHM OPTIMIZATION

The experimental result gives 288.367mpy as a corrosion rate. The software gives the minimum corrosion rate value of 188.866mpy from 50 runs as shown in Figure(14). A small difference can be noted as the genetic algorithm is not the exact fit with linear equations so the experimental and software results are too close to each other.

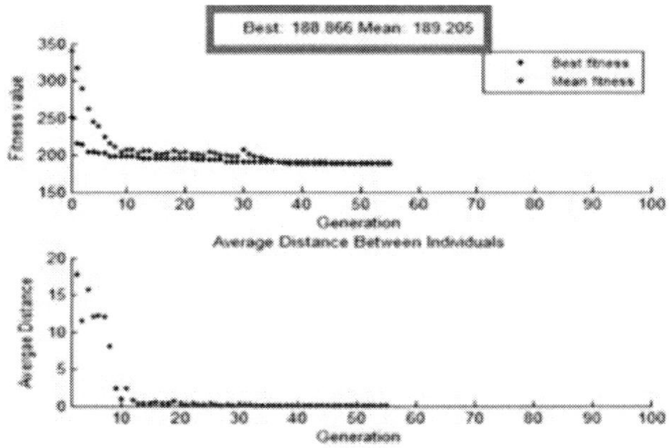

Figure(14): Genetic Algorithm optimization.

CONCLUSION

In this project, all the objectives have been achieved by designing and experimenting.

Firstly; the anticorrosive extract (berries) has been prepared, the prepared extract which contains 4 types of berries namely blueberries, blackberries, cranberries, and raspberries inhibit the mild steel corrosion in sulfuric acid medium with 41.427% efficiency in presence of 5 ml of berries extract.

Secondly, the mild steel corrosion in the presence and absence of berries juice extract were studied where the corrosion rate reaches to 1020.374mpy in absence of berries juice. By using 5 ml of berries juice extract the corrosion rate minimized to 232.022mpy.

Thirdly, the optimum conditions to inhibit the mild steel corrosion were studied by Design Expert V11. At the optimum conditions of 1M of H_2SO_4. 5ml of berries juice extract, 240 minutes and at room temperature the corrosion rate reaches to 232.022mpy as a minimum value.

The actual equation obtained from Design Expert V11 was later optimized using Genetic Algorithm. Genetic Algorithm shows that at 1.001M of H_2SO_4, 15 ml of berries juice extract and 237.208 minutes at room temperature, the corrosion rate is minimized to 188.866mpy.

Finally, mixing tank has been designed with 1.18m height and 2.5m diameter and a turbine agitator has been selected.

ACKNOWLEDGEMENT

I would like to thank god for giving me patience and power to work in my most important part of my life, my father, mother, my brother and my friends for their help and support. I would also like to express my sincere gratitude to Dr. Priy Brat Dwivedi for his support and help in developing my project idea, Mr. M.J.Varghese for his help in RSM analysis, Mr. Hatem Yazidi and Dr. Moammar Ali AlTubi for providing me the technical knowledge in Matlab optimization, Mr.Mohammed Salim Adeeb in providing me the technical help in the laboratory, Ganesh Nagraj Patil, Dr.Lakkim Setty

Rao, and Dr.Santosh Madhav Walke for their help in the calculations of design mixing tank. Last but not the least, I would like to thank National University of Science and Technology-College of Engineering for giving me this opportunity to complete the project in their laboratories and supporting me by providing me with the required chemicals.

REFERENCES

1.Alaneme,K. et al .,2016.Corrosion Inhibitory Properties of Elephant Grass(Pennisetum Purpureum) Extract Effect on Mild Steel Corrosion in 1M HCl Solution :*Alexandria Engineering Journal*.[e-journal]. 55 (8). p. 1069-1076. Available from: https://www.sciencedirect.com/science/article/pii/S1110016816300242.[Accessed: 8th May 2019].

2.Ambrish,S. et al .,2012. The Effect of Environmentally Bening Fruit Extract of Shahjan(Moringa Olifera) on the Corrosion of Mild Steel in Hydrochloric Acid Solution:*Journal of Chemical Engineering Communications*.[e-journal]. 199 (6). p. 63-77. Available from:https://www.tandfonline.com/doi/abs/10.1080/00986445.2011.570390.[Accessed: 6th May 2019].

3.Blumberg,J.,et al.,2013.Carnberries and Their Bioactive Constituents in Human Health:*Journal of Advances in Nutrition*.[e-journal]. 12 (7). p. 618-632. Available from: https://academic.oup.com/advances/article/4/6/618/4595566.[Accessed: 1th May 2019].

4.Bobinaite,R.,Viskelis,P.&Venskutonis,P.,2015.Chemical Composition of Raspberry Cultivars:*International Journal of Molecular Sciences*.[e-journal]. 16 (8). p. 18642-18663. Available from: https://www.researchgate.net/publication/306158828_Chemical_Composition_of_Raspberry_Rubus_spp_Cultivars.[Accessed: 1th May 2019].

5.Cafferty,E.,2010.*Introduction to Corrosion Science*.United States:Springer.
Roberge,P.,2012.*Handbook of Corrosion Engineering*.2nd edition.United States:McGraw-Hill company.
6.Dynamixianc,2015.*Optimal Mixer Design* .[Online]. Available from: https://www.dynamixinc.com/optimal-mixer-design. [Accessed: 14th December 2018]

7.Ibryamoval,Y. et al .,2009.Design of Mixers for Chemical and Biochemical Industry: *Trakia Journal of Sciences*.[e-journal]. 8 (6). p. 6-11. Available from: http://tru.uni-sz.bg/tsj/Vol8.Suppl.3.2010/Y.Ibryamova.pdf.[Accessed: 13th December 2018].

8.Kaume, L. et al.,2012. The Blackberry Fruit A Review on Its Composition and Chemistry, Metabolism and Bioavailability and Health Benefits:*Journal of Agricultural and Food Chemistry*.[e-journal]. 37 (7). p. 5716-5727. Available from: https://www.ncbi.nlm.nih.gov/pubmed/22082199.[Accessed: 6th May 2019].

9.Kumar.A,2012. Inhibition of Mild Steel Corrosion in HCl Solution Using Pipali(Piper Longum)Fruit Extract:*Arabian Journal for Science and Engineering*.[e-journal]. 38 (9). p. 85-97. Available from: https://link.springer.com/article/10.1007/s13369-012-0409-9.[Accessed: 9th May 2019].

10. Michalska,A.& Lysiak.G .,2015.Bioactive Compounds of Bluberries:*International Journal of Molecular Sciences*.[e-journal]. 16 (8). p. 18642-18663. Available from: https://www.ncbi.nlm.nih.gov/pmc/articles/PMC4581264/.[Accessed: 1th May 2019].

11.*Mixing Tank*,2018.[Online]. Available from: https://www.conceptdraw.com/examples/agitator-tank-symbol.[Accessed: 19th November 2018].

12. Mycourses,2016. *Response Surface Modelling-RSM*.[Online]. Available from:https://mycourses.aalto.fi/pluginfile.php/364340/mod_resource/content/1/BOS_Lecture3_2016.[Accessed: 4th May 2019].

13. Ostovari, L. et al.,2009.Corrosion Inhibition of Mild Steel in 1M HCl Soluition by henna extract:*Journal of Corrosion Science*.[e-journal]. 32 (9). p. 1935-1949. Available from: https://www.sciencedirect.com/science/article/pii/S0010938X09002157. [Accessed: 6th May 2019].

14. Petropedia,2018. *Mixing Tank*.[Online]. Available from: https://www.petropedia.com/definition/7742/mixing-tank.[Accessed: 11th December 2018].

15.Raja,P. et al.,2008.Inhibitive Effect of Black Pepper Extract on the Sulphuric Acid Corrosion of Mild Steel:*Journal of Materials Letters*.[e-journal]. 62 (9). p. 1977-2979. Available from: https://www.sciencedirect.com/science/article/abs/pii/S0167577X08001316.[Accessed: 8th May 2019].

16.Sinnot,R.&Towler,G.,2009.*Chemical Engineering Design*.5th edition.India:Elsevier.

17.Xianghong,L. et al.,2009.Inhibition Effect of 6-benzylaminopurine on the Corrosion of Cold Rolled Ssteel in H_2SO_4 Soluition:*Journal of Corrosion Science*.[e-journal]. 51 (9). p. 620-634. Available from: https://www.sciencedirect.com/science/article/pii/S0010938X09000080. [Accessed: 6th May 2019].

590434 Verification of Firefighters' Heuristics and Predictive Model Development through Big Data Analysis

Abstract:

The heuristics accumulated in the field activities of firefighters are reviewed through big data analysis of fire occurrences and emergency responses in Gyeonggi-do Province, Korea and expanded as predictive models to be utilized for proper fire prevention activities according to time, day, and target through quantitative modeling. Empirical rules with high sympathy are collected through direct interviews with firefighters. Among these, three types of high-priority rules are compared and analyzed for big data such as fire occurrence and damage history in Gyeonggi-do in 2018. Big data comparison analyses are conducted, including the number of fires and damages that occurred in Gyeonggi-do in 2018. Furthermore, fire occurrence patterns by region, day of the week, time of day, and building type are derived. In addition to using empirical rules that have been validated through research, relatively inexperienced firefighters also can make better decisions by relying on refined quantitative predictive models replacing empirical rules, including local government and time-based factors that reflect big fire occurrence data.

CONVERTING SHELTER-IN-PLACE LOCATIONS TO ZERO VULNERABILITY SAFE HAVENS

J. A. Rau
MineARC Systems America, LLC
4850 W. Ledbetter Drive
Dallas, Texas 75236
USA

A. Young
MineARC Systems, Pty Ltd
274 Welshpool Rd
Welshpool, WA 6106
Australia

M. McDermott
High Performance Building Solutions
7415 Folger Dr.
Charlotte, NC 28226
USA

CONVERTING SHELTER-IN-PLACE LOCATIONS TO ZERO VULNERABILITY SAFE HAVENS

ABSTRACT

While it is not uncommon to find shelter-in-place or safe haven locations designated across chemical facilities, what varies is the level of protection offered at these sites. Functionality can range from merely a muster point (shelter-in-place) to a positively pressurized room (safe haven). Either way, when determining the room's acceptable performance criteria, the industry generally applies ERPG-3[1] or a vulnerability factor. An acceptable vulnerability factor is often deemed the probability of a fatality being less than one percent.

In 2017, MineARC Systems and High Performance Building Solutions (HPBS) began working with a leading energy company to convert five shelter-in-place spaces to safe havens, which included additional life support systems that would ensure zero percent vulnerability. The objective of this paper is to examine the process to harden the space against contaminated air intrusion and what critical life support systems are essential for converting a shelter-in-place location to a safe haven.

INTRODUCTION

Emergency shelter-in-place (SIP) forms an integral part of any plant's broader emergency response plan. Explosions, fires, and the release of smoke or other toxic gases are some of the types of incidents that can occur despite high levels of planning and safety precautions[2].

Numerous government agencies and other organizations recommend shelter-in-place as a measure to reduce harm in the event of a chemical release. Any building may be used as a temporary measure to reduce health risk from exposure to the toxic materials by simply closing windows and doors and turning off ventilation fans and air conditioning/heating systems.

Three factors govern the effectiveness of a SIP location: 1) the tightness of the building or leakage rate, 2) the concentration of the toxic gas, and 3) the release duration. The tightness of the building is primarily based on building construction; however, temperature differential and wind speed also play a factor. Identification of the maximum toxic concentration occurs during a Facility Siting Study (FSS). FSSs are a hazard analysis that defines Maximum Credible Event (MCE) scenarios using a consequence-based approach to produce a quantitative risk assessment. Determining potential consequences is a necessary step in the process of developing a comprehensive safety plan. The FSS determines the expected release duration based on stored inventories of toxic materials. The higher the toxic concentration and longer the release duration, the less effective sheltering in place becomes.

While occupants may feel safe sheltering-in-place from the external toxic hazard, typical buildings have high leakage rates. A typical wood framed constructed building can have 3-5 air changes per hour (ACH50[3]), rendering it ineffective against most chemical releases. Additionally, sealing to an acceptable airtightness can be time consuming and expensive and even with professional air sealing it is very difficult to air seal existing structures to an acceptable tightness.

A case study for the conversion of five shelter-in-place spaces to zero percent vulnerability safe havens is presented. There are considerations for effectively sealing existing infrastructure and ensuring positive pressure to ensure contaminants do not infiltrate safe haven locations along with the critical element of life support. The critical elements of life support that are of primary importance to human survival in a safe haven are:

- Positive pressure
- Breathable air supply
- Supplemental O_2
- Removal of accumulated CO_2

[1] Emergency Response Planning Guidelines (ERPGs) are values developed by the AIHA Guideline Foundation's Emergency Response Planning Committee. ERPG-3 is the maximum airborne concentration below which it is believed nearly all individuals could be exposed for up to 1 hour without experiencing or developing life-threatening health effects.

2 The CSB is an independent federal agency charged with investigating industrial chemical accidents. Over the last 10 years they have completed investigations into 15 separate chemical release incidents.

[3] ACH50 is the number of times the air volume in a building changes per hour at 50 Pascal of pressure. Commercial air leakage rates are commonly reported at 75 Pascals, but this leakage study was conducted at 50 Pascals based on the targets included in the FSS report.

- Cooling and dehumidifying

CASE STUDY

In 2017 MineARC Systems and its project partner HPBS began working alongside a major energy company to convert five shelter-in-place locations into zero percent vulnerability safe havens.

The process started with the energy company completing its FSS, which included plume modeling to define the chemical release concentrations at each building location along with ideal shelter-in-place locations[4] and internal minimum positive pressure[5]. This information was used by MineARC to calculate indoor concentrations and occupant vulnerability to highlight that shelter-in-place alone would result in the vulnerability of greater than zero percent.

From there, HPBS conducted air infiltration testing following ASTM E779-10 to determine an air leakage baseline (see Table 6) for rooms that were designated to be sealed (pre-weatherization). A site assessment was then completed to review each location and determine its suitability for sealing and conversion to a safe haven. Firstly, the measured baseline infiltration rates were compared to the FEMA 453 estimates based on construction type[6] (see Figure 2). All surfaces of the room were examined for ease of conventional air sealing and mechanical equipment sealing. Rooms that included false floors, large amounts of electrical conduit, cabinetry, computer servers or windows etc. were deemed difficult to seal (i.e. control rooms). Secondly, the building or rooms open floor space was checked to ensure it could accommodate the specified number of people in an emergency. FEMA 453 section 1.6 recommends $20ft^2$ per person for durations of less than 24 hours. With the FSS recommending 4 hours, it was deemed acceptable to have an unrestricted floor space of $6ft^2$ per person.

Manual weatherization (air sealing) of the rooms included installing a cleanroom suspended ceiling system with gaskets and tiles that met the air permeance rate for an air barrier material, cleanroom airtight light fixtures with gaskets, sealing around electrical and plumbing, air sealing all visible air pathways to make sure the interior gypsum board was functioning as the air barrier, installing low-leakage HVAC dampers, installing an emergency shut off for the mechanical system and adding door weather-stripping and sweeps. Post-weatherization air leakage tests were conducted to determine the air infiltration rate of the SIP spaces after completion of the manual air sealing, cleanroom suspended ceiling system and HVAC work. The weatherization air sealing methods improved the air barrier integrity for all the proposed SIP spaces by between 57% and 65% (refer to Table 6).

Aerosol-based envelope sealing was used to reduce leakage rates even further. AeroBarrier® is a building envelope sealing technology that simultaneously measures and seals buildings. AeroBarrier® works by forcing an atomized sealant into orifices under pressure gradually sealing leaks. Rooms were sealed during night shift with personnel able to re-access the rooms by morning. The AeroBarrier® air sealing methods improved the air leakage for all the SIP spaces between 7% and 88% over the post manual weatherization leakage rates (refer to Table 7). AeroBarrier® is not effective for sealing large holes so manual air sealing is required for certain locations prior to using AeroBarrier®.

MineARC engineers once again calculated infiltration concentration rates along with the vulnerability rate. As the probability of a fatality was still higher than zero, MineARC then specified its AirBANK Pressure System and necessary carbon dioxide and oxygen control equipment.

SAFE HAVEN LIFE SUPPORT EQUIPMENT

Primary Life Supporting Atmosphere

A safe haven needs to provide a reliable breathable atmosphere independent of external toxic gas concentrations. Human breathing consumes oxygen (O_2) and expels carbon dioxide (CO_2), carbon monoxide (CO), water vapor, and heat. At complete rest, an average person exhales air at 17% O_2 and 3.2% CO_2. This breathing rate results in a 'resting' O_2 uptake of 0.3L/min of O_2 and a resting CO_2 discharge of 0.24L/min (Brake, Fellows, & Bates, 1999). Taking into consideration the above values, a person in a closed system such as a shelter-in-place succumbs to CO_2 poisoning well before O_2

4 Shelter-in-place location is beyond the scope of this paper, but should be determined based on the chemical hazard, facility layout, and location of personnel.

5 Positive pressure requirements are based on average wind speed at the facility. As wind speed increases so does the SIP leakage rate.

6 The FEMA 453 estimates are intended to estimate filter sizing when ASTM E779 air leakage testing is not conducted.

asphyxiation. Venter et al. (1998) found an average oxygen consumption for 12 entrapped individuals of 0.44 liters per minute, at STP, over 24 hours (including sleeping). A figure of 0.5 liters per minute over a non-sleeping entrapment period of 12 hours is a prudent, conservative design value. This corresponds to a metabolic rate of about 160W or $80W/m^2$ for a typical person with a body skin area of about $2m^2$, which is equivalent to a breathing rate of about 12.5 liters per minute of fresh air.

Consequently, life support systems are needed within a safe haven that can ensure positive pressure keeps contaminants outside, along with replacing the O_2 consumed, and removing exhaled CO_2, moisture, and heat to maintain the atmosphere within acceptable threshold limits.

Positive Pressure

Positive pressure is integral to prevent the ingress of airborne contaminants that may compromise the breathable atmosphere for occupants. Establishing a Class 1 Safe Room (FEMA 453) requires the sufficient release of filtered air from an external source, to form a positively pressurized zone.

Buildings are typically constructed to provide natural or forced ventilation to allow permanent occupancy and potentially combined with large exposed windows and surfaces which can present challenges for ensuring protection during a chemical release. When determining adequate positive pressure within a space, several structural factors need to be considered, including:

- Shelter volume
- HVAC (Heating, Ventilating, and Air Conditioning)
- Door and window tightness
- Existing opening such as vents and elevator shafts
- Indoor and outdoor temperature
- Wind speed and building orientation
- The external screening of the structure

Building design and multiple ingress points suggest the conversion of a smaller interior room within a building can provide a higher level of protection, than the whole building, as it is less subject to wind and buoyancy forces which induce the infiltration of external hazards.

Breathable air from compressed air cylinders venting directly into the SIP forms a positive internal pressure. Determining the amount of air required to reach the necessary level of positive pressure involves measuring the areas internal volume and the relationship between the internal and external atmospheric conditions.

After reaching required ACH (air changes per hour) or tightness, a positive-pressure management system must be sized for the expected shelter duration.

Breathable Air Supply

If the shelter-in-place is airtight, a supply of breathable air is necessary to support occupants over the required shelter duration. Unventilated SIPs that are tightly sealed cannot be occupied for long periods without the risk of high carbon dioxide levels. A secondary source of breathable air must be circulated within the shelter to meet occupancy over the required amount of time.

Clean, breathable air is supplied to a chemical, biological and radiological (CBR) safe room in a few ways: by trapping air inside the room and minimizing the air exchange, passing contaminated air through a filter (not practical for high concentration chemical releases or for chemicals that cannot be filtered) to purify it as it is supplied to the room, and the supplemental supply of oxygen through high pressure compressed air cylinders.

The air supply must contain enough oxygen in its mix, and not contain contaminants like carbon monoxide, carbon dioxide, water, or any other toxins that poison safe haven occupants during entrapment.

Table 1 lists international standards for flowrates for breathable compressed air for self-contained breathing apparatus (SCBA), supplied air respirators (SAR) and refuge chambers.

Table 1 – Standards for acceptable air supply requirements

Air supply requirements	Relevant Standard
Open-circuit positive-pressure SCBA, including all components and accessories except the air cylinder (shell) must resist the permeation and penetration of HD and GB chemical agents when tested for a period of 6 hours on an upper-torso manikin connected to a breathing machine operating at an airflow rate of 40 liters per minute (lpm), 36 respirations per minute, 1.1 liters tidal volume.	NIOSH 42 CFR Part 84 Respiratory Protective Devices
The flow rate should be at least 85L/min (3cfm) per person to remove expired CO2 and CO but at least 255L/min (9cfm) to provide adequate cooling and maintain comfortable conditions. The air inflow is assumed to be at a temperature of 26.5°C and 65% relative humidity	ITA Report No. 14 – Guideline for the provision of refuge chambers in tunnels under construction, May 2014
For manually operated blower type air-hose respirators, the blower shall be capable of delivering air to the space enclosed by the facepiece at a flow rate of at least 85 L/min.	AS/NZS 17:16:2003 Respiratory Protective Devices

The NIOSH standard of 40L/min is for SCBA apparatus (i.e. mouthpiece or face mask) where dilution of CO_2 is not a concern. Conversely, the ITA and Australian standards specify 85L/min and are concerned with both oxygen consumption and carbon dioxide dilution. It is important to note that three times the amount of air is required to maintain comfortable conditions inside a refuge chamber, when compared to simply breathing (Health & Safety Executive UK, 2006).

Unless the positive pressure system is supplying breathable air at a rate that maintains oxygen and compressed gas levels within acceptable ranges (dependent on safe haven occupancy and duration), a CO_2 scrubber and supplementary oxygen supply are required.

Supplemental O_2

O_2 is an essential requirement for cellular activity, and therefore, the absence of O_2 for relatively short periods results in death. Humans within an enclosed environment consume O_2 and if not replaced, would result in an irrespirable atmosphere. An O_2 concentration of less than 11% poses a high risk of death.

Retaining the oxygen levels between acceptable limits is critical. Various international standards outline acceptable O_2 limits for a breathable environment; examples are shown in Table 2.

Table 2 - Standards for allowable O_2 levels

Minimum O_2 Content (% by volume)	Maximum O_2 Content (% by volume)	Relevant Standard
18.5%	23.0%	Mines Safety and Health Administration 30 CFR Parts 7 and 75 Refuge Alternatives for Underground Coal Mines; Final Rule
19.5%	23.5%	Occupational Safety and Health Administration 29 CFR 1910.146 Permit-Required Confined Spaces for General Industry
19.5%	23.5%	Australian Standard AS 2865–2009 Safe working in a confined space

OSHA and Australian Standards for working in a confined space specify an acceptable lower limit of 19.5%, which is consistent with many other standards around the world. The slightly lower MSHA limit of 18.5% is based on a person at rest, not working, therefore, not likely to have trouble in breathing; conditions that are likely to be present in a refuge chamber.

O_2 itself is not flammable but supports and accelerates combustion. OSHA defines O_2 enriched mixtures or atmospheres as those containing more than 23.5% O_2 by volume (OSHA, 2013). In O_2-enriched atmospheres, the reactivity of O_2 significantly increases the risk of ignition and fire.

Within a safe haven, pure O_2 must be utilized in conjunction with a CO_2 removal system.

Removal of Accumulated CO_2

Protection within a safe haven is substantially increased by filtering CO_2 from the air. CO_2 is being produced continuously by the cells of our body and is a byproduct of respiration. The inability to expel this gas or an environment with high concentrations of CO_2 leads to CO_2 poisoning. Breathing air with more than 3% CO_2 can induce dizziness, headaches, and vision disturbance. Prolonged exposure at greater than 6% could result in unconsciousness and death. Listed in Table 3 are examples of international standards of satisfactory CO_2 removal requirements for a breathable environment.

Table 3 – Acceptable CO_2 removal requirements

CO_2 Removal Requirements	**Relevant Standard**
CO_2 removal components shall remove CO_2 at a rate of 1.08 ft^3 (30.5 L)/h/person.	Mines Safety and Health Administration 30 CFR Parts 7 and 75 Refuge Alternatives for Underground Coal Mines; Final Rule
OSHA has established a Permissible Exposure Limit (PEL) for CO2 of 5,000 parts per million (ppm) (0.5% CO2 in air) averaged over an 8-hour workday (time-weighted average or TWA.)	Occupational Safety and Health Administration
5.2.8 Atmospheric monitoring. Atmospheric monitoring There should be the capability to monitor temperature continuously, and for carbon monoxide, carbon dioxide, and oxygen in the chamber during use in both externally supported or standalone modes. The concentration of carbon dioxide should not exceed 10000 ppm.	ITA Report No. 14 – Guideline for the provision of refuge chambers in tunnels under construction, May 2014

The CO_2 that is exhaled by the occupants within the safe haven must be removed by, constant flushing by an external breathable air source (e.g. compressed air cylinders used for positive pressure) or by the regeneration of the internal atmosphere with a CO_2 absorbent (scrubber).

Different regulations around the world, state varied human production rates between 30–50L per hour per person. When considering an acceptable production rate; it must be noted that the rate of breathing is primarily triggered by the CO_2 content of inspired air with higher CO_2 levels triggering faster respiration rates (Brake, Fellows, & Bates, 1999) (i.e. if the maximum allowable CO_2 concentration inside the safe haven is 1%, a lower production rate is acceptable compared to a 3% CO_2 limit).

If CO_2 removal mechanisms are inadequate, the occupants can develop CO_2 toxicity by re-breathing their own exhaled CO_2.

Cooling and Dehumidifying

Thermal tolerability with a safe haven should be addressed during the design phase. Assumptions must be made in preparation for safe haven design in regard to the number of personnel, activity level, available hydration and heat production. The minimum airflow to maintain a comfortable environment is set by taking these assumptions into account as well as the input temperature and humidity of supplied air.

The healthy human body maintains its internal temperature around 37°C. An essential requirement for continued normal body function is that the deep body temperature is maintained within the narrow 1°C limit (Epstein, Moran, 2006). A change of body temperature exceeding 1°C occurs only during illness or when environmental conditions surpass the body's ability to cope with extreme temperatures.

When a person is subject to extreme heat and humidity over some time, the person may experience a very rapid progression of symptoms or exhibit a few of the less significant symptoms before falling victim to more severe forms of heat-related illness.

The acceptable cooling requirements for underground refuge chambers are defined in various international standards, examples of which are listed in the below table.

Table 4 – Standards for acceptable cooling requirements

Cooling Requirements	Relevant Standard
Simulated emergencies, in which a full complement of people has occupied a refuge chamber for a significant period, indicate that humidity and temperature can increase very rapidly to potentially heatstroke-inducing levels. Refrigerated air conditioning is strongly recommended for both externally supported and stand-alone refuge chambers to counter this potentially dangerous problem. Inevitably, this places a heavy demand on the stand-alone power supply, but there are systems available that can cope, and at an acceptable cost.	Department of Mines, Industry Regulations and Safety, 2005 Refuge Chambers in Underground metalliferous mines – guideline: Safety and Health Division, Department of Industry and Resources Western Australia, 32 pp
Apparent temperature in the fully occupied refuge chamber must not exceed 95 °F.	Mines Safety and Health Administration 30 CFR Parts 7 and 75 Refuge Alternatives for Underground Coal Mines; Final Rule
It should be possible to maintain an internal temperature not exceeding 30 °C wet bulb globe temperature for at least 24 hours	ITA Report No. 14 – Guideline for the provision of refuge chambers in tunnels under construction, May 2014

MSHA regulations specify a maximum internal apparent temperature[7] of 95°F (35°C). NIOSH specifies an equivalent maximum wet bulb temperature of 28.9°C (NIOSH, 2007).

The temperature maintained inside the safe haven is determined by the metabolic heat liberated by the occupants, dissipated to the external environment by the heat transfer characteristics of the room or building. The external temperature primarily influences the heat transfer while the humidity is the combination of moisture produced by the occupants (via exhaled air and sweat) and water liberated by CO_2 scrubbing chemicals.

It is important to remember that in most cases, before sheltering-in-place, the air conditioning system will need to be shut down (unless it is a split system). For short durations, artificial cooling is not a critical requirement unless external temperatures are extraordinarily high, and/or the safe haven's volume is small compared to its occupancy.

INFILTRATION TESTING

Background

High Performance Building Solutions Inc. (HPBS) was contracted to perform an air infiltration study for spaces identified as candidates to be used as shelter-in-place (SIP) in the event of an accidental chemical release. This SIP strategy uses the existing structure and its indoor atmosphere to separate individuals from a hazardous outdoor atmosphere temporarily.

A series of pre-weatherization air leakage tests were conducted to determine the baseline air infiltration rate of potential SIP spaces. Manual air sealing, a cleanroom suspended ceiling system and low-leakage HVAC dampers were installed to reduce the overall air infiltration of the SIP spaces and to close off air leakage pathways from non-protected spaces in the building through the HVAC system. Post-weatherization air leakage tests were conducted to determine the air infiltration rate of the SIP space after the completion of weatherization and HVAC work. The SIP spaces were additionally air sealed using the proprietary AeroBarrier® system to further reduce infiltration rates.

Scope

Air sealing work was completed, and low-leakage HVAC dampers were installed to reduce the overall air infiltration of the SIP spaces and to isolate the HVAC from non-protected spaces in the building. Post-weatherization air leakage tests were conducted to determine the air infiltration rate of the SIP space after the completion of air sealing, suspended ceiling and HVAC work. AeroBarrier® air sealing was conducted on SIP spaces as a pilot project on September 18 through September 21, 2017. Post AeroBarrier® air leakage tests were conducted on September 21 and September 22, 2017, to determine the air infiltration rate of the SIP space after the completion of the AeroBarrier® sealing.

7 Apparent temperature is the temperature equivalent perceived by humans, caused by the combined effects of air temperature and relative humidity.

Following ASTM E 779-10, pressurization air leakage tests were conducted on the following spaces recommended in the Facility Siting Study (FSS).

Table 5: SIP specifications

SIP Space	Occupancy	Floor Area (Square Feet)	Volume (Cubic Feet)
SIP Space 1	40	990	10,481
SIP Space 2	6	955	8,210
SIP Space 3	5	391	3,917
SIP Space 4	10	1,697	16,968
SIP Space 5	40	293	2,637

Air Leakage Test Results

Positive pressure air leakage test results are reported in four different ways:

CFM4 is the quantitative volume of air passing through an enclosure under a pressure difference of 4 Pascals (equivalent to a 5mph wind). This is a measurable physical quantity and the value recommended in the FSS to eliminate toxic gas ingress into the SIP spaces.

CFM25 is the quantitative volume of air passing through an enclosure under a pressure difference of 25 Pascals (0.1-in w.g.). FEMA 453 requires Class 1 Safe Havens be pressurized to 25 Pascals. This is a measurable physical property.

CFM50 is the quantitative volume of air passing through an enclosure under a pressure difference of 50 Pascals (0.2-in w.g.).

ACH50 is the quantitative volume of air passing through an enclosure under a pressure difference of 50 Pascals (0.2-in w.g.) normalized by the enclosure volume. ACH is the abbreviation for air changes per hour. It is the number of times the air volume in a building changes per hour at 50 Pascals of pressure.

Table 6 below shows the CFM50 and ACH50 for each of the SIP spaces pre-weatherization and post-weatherization along with the percentage reduction in leakage. SIP Space 1 and SIP Space 5 did not have pre-weatherization results as initially the whole building was tested. Smaller spaces were selected as the whole building was too large to feasibly convert to a SIP space (re-testing of the new SIP spaces was not completed). The control room, SIP space 4, was abandoned in favor of a smaller adjacent room due to high leakage post manual weatherization and the computers that could not be shut down, which prevented the use of AeroBarrier® in the space.

Table 6: Post-weatherization leakage results

Safe Haven Space	CFM50		ACH50		% Reduction
	Pre-Weatherization	Post-Weatherization	Pre-Weatherization	Post-Weatherization	
SIP Space 1	N/A	1,871	N/A	10.7	N/A
SIP Space 2	3,623	1,560	26	11.4	57%
SIP Space 3	385	135	6	2.1	65%
SIP Space 4	N/A	N/A	N/A	N/A	N/A
SIP Space 5	N/A	655	N/A	19.4	N/A

The AeroBarrier® air sealing was prescribed a target. According to the AeroBarrier® simultaneous air leakage calculations, all the spaces met and surpassed this airtightness target; AeroBarrier® sealing was halted at this point. Table 7 shows the CFM50 and ACH50 for each of the SIP spaces as tested by HPBS post AeroBarrier® along with the percentage reduction. The HPBS results were utilized to calculate the amount of positive pressure required to convert the rooms to positive pressure safe havens.

Table 7: Post AeroBarrier® leakage results

Safe Haven Space	CFM50		ACH50		% Reduction
	Post-Weatherization	Post AeroBarrier®	Post-Weatherization	Post AeroBarrier®	
SIP Space 1	1,871	455	10.7	2.6	76%
SIP Space 2	1,560	182	11.4	1.3	88%
SIP Space 3	135	125	2.1	1.9	7%
SIP Space 4	N/A	173	N/A	3.3	N/A
SIP Space 5	655	138	19.4	3.1	79%

The AeroBarrier® air sealing methods improved the air barrier integrity for all the proposed SIP spaces between 7% and 88% over the post manual weatherization leakage rates. Air leakage was still observed behind many electrical penetrations including HVAC thermostats, and fire and smoke alarms. These were covered with plastic to protect them from the AeroBarrier® Sealant but had hidden holes behind them.

Figure 1 displays the post AeroBarrier® leakage rate in comparison to the FEMA 453 estimates for air leakage rates based on construction types. FEMA 453 estimates are reported in square feet of floor area at 25 Pascals and do not take into consideration the ceiling height and volume of the space.

Figure 1: Post AeroBarrier® results compared to FEMA 453 estimates

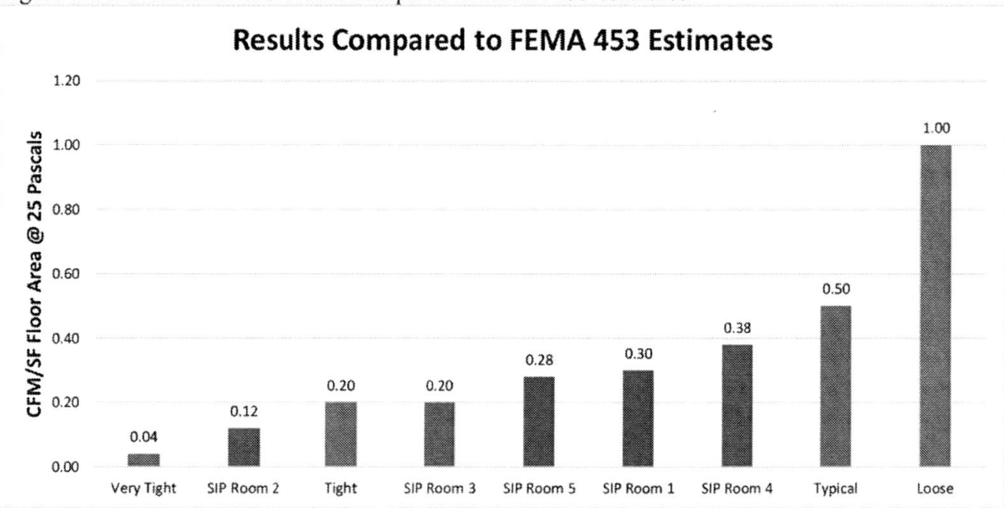

Figure 2 is an estimated leakage rate based on construction type to estimate the filtration and positive pressure system size under the FEMA 453 standard. All the SIP locations were between tight and typical except for SIP Room 2, which was a small prefabricated metal constructed portable building.

Figure 2: Estimated leakage rate per square foot of floor space to achieve 0.1 iwg (25Pa) overpressure

Construction Type	cfm per square foot of floor area
Very tight: 26-inch thick concrete walls and roof with no windows	0.04
Tight: 12-inch thick concrete or block walls and roof with tight windows and multiple, sealed penetrations	0.20
Typical: 12-inch thick concrete or block walls with gypsum wall board ceilings or composition roof and multiple, sealed penetrations	0.50
Loose: Wood-frame construction without special sealing measures	1.00

Table 8 shows the combined AeroBarrier®, and manual air sealing methods improved the airtightness for all the proposed SIP spaces between 68% and 95% over the original leakage rates.

Table 8: Air infiltration improvement for SIP spaces

Safe Haven Space	ACH50			% Reduction
	Pre-Weatherization	Post-Weatherization	Post AeroBarrier®	
SIP Space 1	N/A	10.7	2.6	76%
SIP Space 2	26.5	11.4	1.3	95%
SIP Space 3	5.9	2.1	1.9	68%
SIP Space 4	N/A	N/A	3.3	N/A
SIP Space 5	N/A	19.4	3.1	84%

Figure 3 compares the ACH50 pre-weatherization, post-weatherization and AeroBarrier® air infiltration results.

Figure 3: Pre-weatherization, post-weatherization and AeroBarrier® comparison

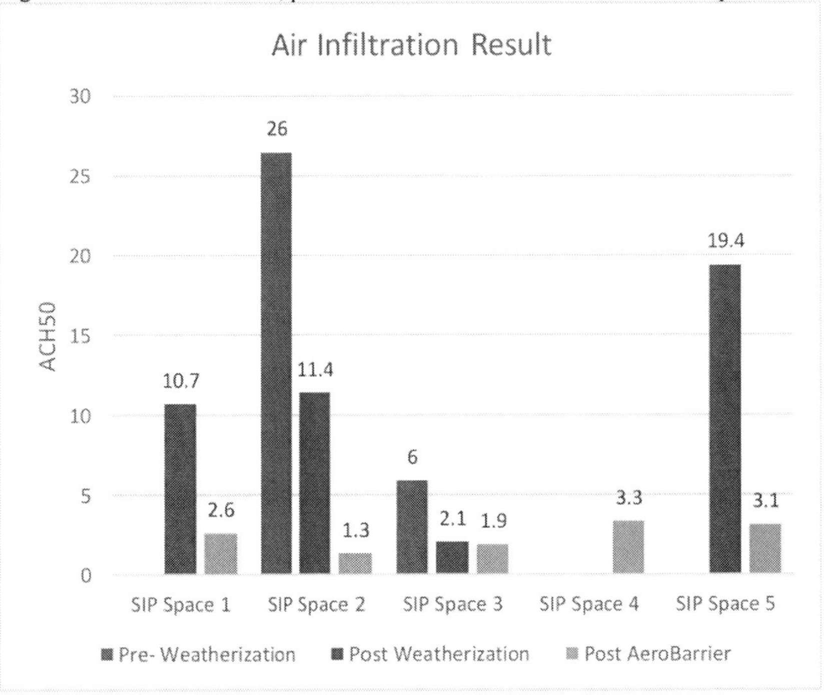

Life Support Systems

MineARC re-calculated the vulnerability for each SIP space based on the post AeroBarrier® leakage rates. All SIP spaces still had a vulnerability of 100% and therefore required the installation of independent life support systems to be deemed zero percent vulnerability safe havens.

Figure 4 shows the toxic gas concentration, duration of the release, and vulnerability for each SIP space. As the largest MCE was a full tank rupture the toxic concentrations were extremely high, but the duration was short at one minute.

Figure 4: SIP space vulnerability post AeroBarrier®

Safe Haven Location	TOXIC GAS				
	Worst Case Threat Scenario	Toxic External [ppm]	Duration of Reslease [min]	ACM	Vulnerability
SIP Space 1	Tank Rupture	585,000	1	0.032	100%
SIP Space 2	Tank Rupture	652,000	1	0.042	100%
SIP Space 3	Tank Rupture	660,000	1	0.007	100%
SIP Space 4	Tank Rupture	412,000	1	0.042	100%
SIP Space 5	Tank Rupture	600,000	1	0.028	100%

The leakage rate at 4 Pascals of pressure difference is slightly more than would occur under a 5 mile per hour wind used in the dispersion model in the FSS report based on climatic data for the facility (to maintain positive pressure and neutralize air infiltration). Post-weatherization infiltration rates in cubic feet per minute at 4 Pascals of pressure for each of the SIP spaces is shown in Table 9.

Table 9: Post AeroBarrier® air leakage results at CFM4

Safe Haven Space	Volume (Cubic Feet)	CFM4
SIP Space 1	10,481	99
SIP Space 2	8,210	32
SIP Space 3	3,917	24
SIP Space 4	3,110	45
SIP Space 5	2,637	21

The air leakage test results provide a mathematical formula that can be used to calculate the amount of breathable compressed air cylinders (437ft^3) that must be added to each candidate SIP space to maintain a 4 Pascal positive pressure to overcome a 5mph design wind coincident with a toxic release.

MineARC Systems was contracted to install life support systems to convert the SIP spaces to positive pressure safe havens. The MineARC AirBANK Positive Pressure System provides rapid pressurization, which is activated and maintained using the AirBANK Control via a simple human-machine interface (HMI) touch screen. MineARC's integrated Aura-FX Gas Monitor ensures breathable air automatically remains within acceptable limits. Additionally, the installation of supplementary oxygen and carbon dioxide scrubbing systems in safe havens, are as needed.

Figure 5 shows the number of 437 cubic feet cylinders required for each safe haven location. The minimum breathable air per person is 2.5cfm (70L/min) per person based on MineARC's experience and testing within the refuge chamber industry.

Figure 5: Quantity of cylinders required for 4Pa positive pressure

	SAFE HAVEN SPECIFICATIONS		POSITIVE PRESSURE (@ 4Pa)						
Safe Haven Location	Occupancy	Duration	CFM	CFM per Floor Space	ACH	No. Cylinders	No. Cylinder Racks	Breathable Air per Person [CFM]	Breathable Air per Person [LPM]
Safe Haven Space 1	40	4	99	0.10	0.6	55	5	2.5	70.8
Safe Haven Space 2	6	4	32	0.03	0.2	18	2	5.4	152.9
Safe Haven Space 3	5	4	24	0.06	0.4	14	2	4.8	135.9
Safe Haven Space 4	10	4	45	0.14	0.9	25	3	4.5	127.4
Safe Haven Space 5	40	4	21	0.07	0.5	12	2	0.6	17.0

The maximum CO_2 concentration is at 1% and O_2 between 19.5% and 23.5% based on OSHA limits (refer to Table 2). Due to the volume and number of people in Safe Haven Space 5, it required supplemental oxygen and carbon dioxide scrubbing to maintain the internal gas concentrations within specified limits. Three sodium chlorate oxygen candles and two AirGEN CO_2 scrubbers were included in the design.

Figure 6: Internal gas concentrations in safe havens

	SAFE HAVEN SPECIFICATIONS		POSITIVE PRESSURE (@ 4Pa)			O2 & CO2 CONCENTRATIONS (@ 4Pa)			
Safe Haven Location	Occupancy	Duration	CFM	CFM per Floor Space	ACH	Indoor O_2 %	No. O_2 Candles	Indoor CO_2 %	No. AirGEN Scrubbers
Safe Haven Space 1	40	4	99	0.10	0.6	20.45	0	0.50	0
Safe Haven Space 2	6	4	32	0.03	0.2	20.84	0	0.17	0
Safe Haven Space 3	5	4	24	0.06	0.4	20.77	0	0.23	0
Safe Haven Space 4	10	4	45	0.14	0.9	20.66	0	0.31	0
Safe Haven Space 5	40	4	21	0.07	0.5	18.67	3	2.04	2

In 2018 MineARC completed installation and training on the five safe haven spaces. Each safe haven was tested during the commissioning process to ensure it could maintain acceptable positive pressure.

CONCLUSIONS

The conversion of shelter-in-place (SIP) locations to zero percent vulnerability safe havens is a four-step process consisting of the following:

1) Identifying the correct SIP location based on ease of sealing and small volume based on the required occupancy.
2) Manual air sealing of SIP spaces.
3) AeroBarrier® sealing of SIP spaces.
4) Installation of life support equipment.

Identifying SIP spaces that can seal from the outset is crucial to a successful result. In the case study, three of the five original locations were abandoned in favor of rooms in adjacent locations due to high leakage rates or reducing the size to make the project more cost effective. Control rooms pose the most difficulty in trying to seal due to the false floor and a myriad of cable tray and pass-throughs installed for the computer systems. The smaller the room and the more robust the construction (i.e. prefabricated metal or concrete buildings), the better it seals.

Manual air sealing needs to be overseen by someone with familiarity in building tightness. While most of the tasks are general construction, standard contractors are not trained to work at the level of precision needed to ensure proper sealing. Even a tiny orifice or ceiling tile not sealed correctly can ruin a leakage test result. HPBS and MineARC Systems have developed a system for ensuring airtightness in construction methodology and by avoiding un-sealable spaces or spaces that are not cost effective to seal.

For future construction projects, incorporation of SIP spaces should occur during the design phase. Air barrier commissioning is standard in new construction and should include the following tasks at a minimum: design review of the plans with specifications for the air barrier and mechanical system - specifically as these relate to spaces designated for use as safe havens; quality assurance inspections during construction to ensure a continuous air barrier is in place; and final air leakage testing of the continuous air barrier to ensure the leakage rate of the safe haven spaces are within acceptable limits.

Complementary life support equipment is essential to sustain life within an airtight environment. For personnel to remain within the sealed safe haven, the environment must be habitable. Air is to be free of toxins such as carbon dioxide and acceptable levels of oxygen maintained. Positive pressure systems reduce the ingress of toxins and provide some or all of the required source of breathable air. Additional equipment such as compressed air, oxygen cylinders and air scrubbers are used to meet the shortfall. Furthermore, internal conditions must consider the human condition, supporting workable body temperature and humidity.

Reducing the risk of harm to personnel is a priority for emergency response planning. While the likelihood of an MCE occurring may be low, the risk to on-site personnel in the event of an incident is high. Converting shelter-in-place (SIP) locations to zero vulnerability safe havens improves emergency response.

REFERENCES

ASTM E779-10, *Standard Test Method for Determining Air Leakage Rate by Fan Pressurization*, ASTM International, West Conshohocken, PA, 2019, www.astm.org

Brake, Fellow, & Bates (1999). *Criteria for the design of emergency refuge stations for an underground metal mine*. Aus. IMM Proceedings, No.2, 1999, pp 1–7.

Department of Mines and Petroleum, 2013, *Refuge chambers in underground mines — guideline: Resources Safety*, Department of Mines and Petroleum, Western Australia, 41 pp.

Epstein, Y., & Moran D. (2006). *Thermal Comfort and the Heat Stress Indices,* Industrial Health, 44, 388–398.

FEMA 453. (2006). *Design Guidance for Shelters and Safe Rooms.* Federal Emergency Management Agency

Health & Safety Executive UK (2008). *Guidance and Information on the role and design of safe havens in arrangements for escape from mines.* Retrieved from http://www.hse.gov.uk/pubns/mines08.pdf

ITA Working Group 5 (2016). *Provisions for refuge chambers in tunnel projects under construction.* ITA Report N°014 - V2 / March 2018

Mines Safety and Health Administration (2008). *30 CFR Parts 7 and 75 Refuge Alternatives for Underground Coal Mines; Final Rule.*

National Institute for Occupational Safety and Health (2007). *Research Report on Refuge Alternatives for Underground Coal Mines.* Retrieved from http://www.cdc.gov/niosh/docket/archive/pdfs/NIOSH-125/125 ResearchReportonRefugeAlternatives.pdf

Occupational Safety & Health Administration (2013). *Permit-required confined spaces.* Retrieved from http://www.osha.gov/pls/oshaweb/owadisp.show_document?p_table=STANDARDS&p_id=9797

Suming Du. (2014). *On the effectiveness of shelter-in-place as a measure to reduce harm from atmospheric releases. Journal of Emergency Management, Vol. 12, No, 3, May/June 2014.*

Venter. J. (1998). *Portable Refuge Chambers: Air or Tomb in Underground Escape Strategies.* School for Mechanical and Materials Engineering, Potchefstoom University, Republic of South Africa

Worksafe Australia National Standard (2005). *Australian Standard AS: NZS 1716:2003 Respiratory protective devices.*

Yantosik. G. (2006). *Shelter-in-Place Protective Action Guide.* The Chemical Stockpile Emergency Preparedness Program (CSEPP). U.S. Department of Energy

589008 How the Field of Human Factors Helps Solve the Riddle of Serious Injuries and Fatalities

Abstract:

Process and occupational safety are often treated as distinct endeavors, with the former tending to focus on facility system function and the latter on individual worker wellbeing. Though other differences certainly exist between the two, both endeavors are generally purposed to prevent incidents that can cost lives, health, time, and money, in addition to employee morale and public trust. The overarching thesis of this paper is that the field of Human Factors is uniquely positioned to effect positive change in both of these major safety categories.

The backdrop for our argument is the recent finding that while injury and illness incidence rates published by the Bureau of Labor statistics have decreased in recent years, the occurrences of serious injuries and fatalities have declined at a significantly slower rate. We discuss possible explanations for this finding but note that regardless of the preferred explanation, the trend should motivate scientists, engineers, and other safety professionals to revisit long-held accident causation theories holding that near misses – these are incidents that could have resulted in serious injuries and fatalities but did not – are precursors of future serious incidents. While this idea still has merit, the just-mentioned statistics warrant reconsideration of the relations between serious and less-serious incidents to explain why the rates of the two are not decreasing in lockstep. We suggest these data indicate additional (human) factors are involved.

To address this riddle and to support our general thesis, we (a) review core tenets of process safety, occupational safety, and the field of Human Factors, (b) point out that the human is always an element common to both process and occupational safety contexts, and (c) discuss how preventing or mitigating human error can benefit both. We offer specific examples from both safety contexts to show how Human Factors perspectives and techniques can help. Examples touch on standards, policies, and other administrative controls (e.g., procedures), human-machine interfaces of both workplace hardware (e.g., valves) and software (e.g., SCADA systems), environments (e.g., physical, regulatory, and cultural), and communications between personnel (e.g., elements of team resource management). We conclude that fundamental Human Factors considerations can help prevent and mitigate human performance gaps and therefore increase the rate of reduction in serious injuries and fatalities in both process and occupational safety.

589233 Initial Conditions Used in Atmospheric Dispersion Models:Issues with Commonly Used Assumptions for Gas Jets and Plumes

Abstract:

When simulating jets and plumes, all atmospheric dispersion models require initial conditions that appropriately parameterize the behavior of a gas release to the atmosphere. Often, process conditions are such that loss of primary containment of a gas will result in choked flow conditions. A great deal of experimental and theoretical consideration has been given to the flow from stagnation conditions to choked flow conditions, but less experimental work has centered on describing the flow through the depressurization process to atmospheric pressure where atmospheric dispersion models would be applicable. Traditional modeling of the depressurization process has taken the simplified approach of assuming a (depressurized) "top hat" profile (laterally constant) with application of the principles of conservation of mass, momentum, and energy. Such assumptions are known to result in predicted depressurized velocities that exceed the (sonic) velocity at the point of loss of containment, and because of the energy balance, the associated temperatures are significantly less than temperatures at the choked conditions. Physically, these effects could possibly self-cancel because as jet velocity slows, the decrease in kinetic energy of the jet will increase the jet temperature. However, no atmospheric dispersion model used for routine hazard assessment purposes accounts for kinetic energy in the energy balance. The traditional modeling approaches of the depressurization process are reconsidered based on analysis of available data, and an alternative approach to modeling the depressurization process is proposed. Example cases showing the important differences between the two approaches are discussed.

Fault Tree Uncertainty Analysis

Raymond "Randy" Freeman
S&PP Consulting
12303 Lake Shore Ridge
Houston, TX 77041
Cell: 713 408 0357
Rafree@yahoo.com

For Presentation at the 2020 Spring National AIChE Meeting
Houston, Texas

March 30, 2020

ABSTRACT

Fault tree analysis (FTA) is a widely used methodology in the process industries. FTA is used for the development of failure mechanisms, computation of failure frequencies and the determination of the probability of failure on demand of safety systems. Much of the data used in a FTA study are uncertain. For example, the failure rate of a pump is often not known with great precision. Likewise the failure rates of instrumentation are often known only within some defined limits. The common practice, used by analysts in the quantification of a fault tree, is to use the most likely or best guess as to the needed failure rate data. The use of best guess values as data inputs to the quantification of a fault tree creates uncertainty in the computed results.

This paper presents a general methodology for the determination of the impact of uncertainty on the results of a fault tree study. The general methodology is based on the mathematics of propagation of error and variance contribution analysis. For a fault tree that contains 10 event minimal cut sets, the methodology will compute the variance in the probability (or frequency) of the top event with a maximum error of 5%. For 5 event minimal cut sets the error in variance of the top event probability (or frequency) will be no more than 2%. An example is presented to illustrate the application of the fault tree uncertainty analysis methodology to a real world problem.

Keywords

Quantitative Risk Assessment, Fault Tree Analysis, Uncertainty Analysis

1. Background

Fault tree analysis is a widely used method for representation of failure mechanisms in chemical plants. The methodology is described in the CCPS CPQRA guideline book [Ref. 1] and is extensively discussed in Chapter 9 of Lees [Ref. 2]. Lacking in these discussions is a systematic method to quantify the uncertainty of the results of the fault tree calculations.

Fault tree analysis has also been used in the aircraft and aerospace industries, nuclear power plant review and is recognized by the OSHA Process Safety Management (PSM) standard [Ref. 3] as a method for process hazards analysis.

2. Review of Error Propagation and Variance Contribution Analysis (VCA) Methodology

The mean and variance of a function of random variables can be approximated using the method described by Haugen [Ref. 4] and applied by Freeman [Ref. 5, 6, 7]. Define an arbitrary function of a set of random variables, x_i, as:

Let

$$Y = F(x_i) \qquad \text{(Eq 1)}$$

The mean of Y can be estimated using the following approximation:

$$E(Y) = F[\,E(x_i)] \qquad \text{(Eq 2)}$$

Where:

$E(Y)$ = expected value of random variable Y = mean of Y

$E(x_i)$ = expected value of random variable x_i = mean of x_i

The variance of Y can likewise be estimated as:

$$V(Y) = \sum_{i=1}^{n} \left[\frac{\partial Y}{\partial x_i}\right]^2 V(x_i) \qquad \text{(Eq 3)}$$

Where:

$V(Y)$ = variance of random variable Y as defined above in Equation 1

$V(x_i)$ = variance of random variable x_i as defined above in Equation 1

Note that the variance is simply the square of the standard deviation. Using the variance will simplify the mathematics that is described below. The contribution of each independent random variable to the overall variance in the function is:

$$V(Y \text{ from } x_i) = \left[\frac{\partial Y}{\partial x_i}\right]^2 V(x_i) \qquad \text{(Eq 4)}$$

The relative contribution of each term to the overall variance V(Y) is a measure of the importance in the uncertainty in the particular random variable, x_i. In effect, this is a sensitivity analysis combined with a uncertainty evaluation. The variance contribution combines the sensitivity in the answer to changes in the uncertain random variable, x_i, with a measure of the uncertainty in the random variable, x_i. The overall variance in Y is found by summing the sensitivity weighted variances from each random variable.

Numerical Estimate of the Sensitivity

Let us return to the fundamental definition of the derivative.

As before, set y = F(xi)

$$\frac{\partial y}{\partial x} = \lim_{\Delta xi \to 0}\left[\frac{F(xi+\Delta xi) - F(xi)}{\Delta xi}\right] \qquad \text{(Eq 5)}$$

We are interested in a numerical estimate of the derivative. Therefore, we will make the following approximation.

$$\frac{\partial y}{\partial x} \sim \left[\frac{F(xi+\Delta xi) - F(xi)}{\Delta xi}\right] \qquad \text{(Eq 6)}$$

Linear Functions

Let us examine the approximation for a simple linear function.

Let Y = AX + B (Eq 7)

What is the sensitivity of Y to a change in variable X? By simple calculus:

$$\frac{\partial y}{\partial x} = A \qquad \text{(Eq 8)}$$

We can also find the sensitivity using the numerical approximation as:

$$\frac{\partial y}{\partial x} \sim \left[\frac{F(xi+\Delta xi) - F(xi)}{\Delta xi}\right] \qquad \text{(Eq 9)}$$

$$\frac{\partial y}{\partial x} \sim \left[\frac{(A(xi+\Delta xi)+B) - (A(xi)+B)}{\Delta xi}\right] \qquad \text{(Eq 10)}$$

$$\frac{\partial y}{\partial x} \sim \left[\frac{(A(xi+\Delta xi)) - (A(xi))}{\Delta xi}\right] \qquad \text{(Eq 11)}$$

$$\frac{\partial y}{\partial x} \sim \left[\frac{((A\,xi+A\,\Delta xi)) - (A(xi))}{\Delta xi}\right] \qquad \text{(Eq 12)}$$

$$\frac{\partial y}{\partial x} \sim \left[\frac{A\,\Delta xi}{\Delta xi}\right] \qquad \text{(Eq 13)}$$

$$\frac{\partial y}{\partial x} = A \qquad \text{(Eq 14)}$$

Note that for a simple linear equation the numerical result for the sensitivity is exactly the same as the analytical expression. There is no restriction on the size of the perturbation (Δxi) used in the calculations. Previously, Freeman [Ref 7] has suggested using a 10% perturbation in xi to do the sensitivity calculations. However any size of perturbation will work and generate the correct sensitivity.

This is a general result. As long as the function can be expressed as a linear function of a number of variables, xi, the sensitivity can be directly calculated by a simple perturbation of the function. The resulting sensitivity will be exactly the same as would be obtained using analytical methods.

Non-Linear Expressions

In the quantification of a fault tree, the top event frequency is computed using failure rates of the component devices and time that measures the interval between system validations. In a minimal cut set of two or more basic events, the time variable may be the same. For example the test interval for a valve may be the same for a pressure transmitter. A minimal cut set involving both the valve and the pressure transmitter would create a 2rd order term in the test interval. If the test interval is uncertain, how do we compute the sensitivity of the top event frequency with respect to the test interval? We will now explore how to deal with this type of problem when completing the uncertainty analysis.

As before, let

$$y = F(xi) = Axi^n + B \qquad \text{(Eq 15)}$$

What is the sensitivity of Y to a change in variable X? By simple calculus:

$$\frac{\partial y}{\partial x} = nAxi^{n-1} \qquad \text{(Eq 16)}$$

Let us evaluate the numerical approximation for this case. If n=2, the function of interest is:

$y = F(xi) = Axi^2 + B$ (Eq 17)

From basic calculus, the exact analytical expression for the sensitivity is:

$$\frac{\partial y}{\partial x} = 2\ A(xi) \qquad \text{(Eq 18)}$$

If xi = 1

$$\frac{\partial y}{\partial x} = 2\ A \qquad \text{(Eq 19)}$$

Again the approximation of the derivative is:

$$\frac{\partial y}{\partial x} \sim \left[\frac{(A(xi+\Delta xi)^2+B)-(Axi+B)}{\Delta xi}\right] \qquad \text{(Eq 20)}$$

$$\frac{\partial y}{\partial x} \sim \left[\frac{F(xi+\Delta xi)-F(xi)}{\Delta xi}\right] \qquad \text{(Eq 21)}$$

Inserting the equation for F(xi) we obtain:

$$\frac{\partial y}{\partial x} \sim \left[\frac{(A(xi+\Delta xi)^2+B)-(Axi+B)}{\Delta xi}\right] \qquad \text{(Eq 22)}$$

$$\frac{\partial y}{\partial x} \sim \left[\frac{(A\ (xi^2+2\Delta xi+\Delta xi^2)\ +B)-(Axi+B)}{\Delta xi}\right] \qquad \text{(Eq 23)}$$

$$\frac{\partial y}{\partial x} \sim \left[\frac{A\ (xi^2+2\Delta xi+\Delta xi^2)\ -Axi}{\Delta xi}\right] \qquad \text{(Eq 24)}$$

$$\frac{\partial y}{\partial x} \sim A\left[\frac{(xi^2+2\Delta xi+\Delta xi^2)\ -xi}{\Delta xi}\right] \qquad \text{(Eq 25)}$$

Setting xi =1, and using a 10% perturbation as $\Delta xi = 0.1\ xi = 0.1$

$$\frac{\partial y}{\partial x} \sim A\left[\frac{(1^2+(2)(0.1)1+0.1^2\ )\ -1}{0.1\ (1)}\right] \qquad \text{(Eq 26)}$$

$$\frac{\partial y}{\partial x} \sim A\left[\frac{(2)(0.1)+0.01}{0.1}\right] \qquad \text{(Eq 27)}$$

$$\frac{\partial y}{\partial x} \sim A\left[\frac{0.21}{0.1}\right] \qquad \text{(Eq 28)}$$

$$\frac{\partial y}{\partial x} = 2.1\ A \qquad \text{(Eq 29)}$$

Note that the numerical estimate (Eq 29) is approximately 5% greater than the analytical value (Eq 19). As shown in Table 1, for higher order equations the error will increase as the size of the perturbation becomes more important. Table 2 shows the impact of reducing the perturbation size to 1%. For 10th order variables (say a minimal cut set with 10 basic events) the error is 5%. For almost all engineering evaluations, a maximum error of 5% should be adequate.

For cases where the test frequency, repair time, or mission time appears in a minimal cut set 3 or more times, I suggest that the following perturbation in time be used to compute the sensitivity of the top event:

$$\Delta time = 0.01\ time \qquad \text{(Eq 30)}$$

The sensitivity of the top event frequency to all uncertain variables can be computed using a perturbation of 1%.

$$\Delta xi = 0.01\ xi \qquad \text{(Eq 31)}$$

3. Fault Tree Analysis

Fault trees are often used to analyze potential failures of an engineered system. A fault tree is a failure logic diagram that shows the relationship of device or system failures leading to an accident. Fault tree analysis is covered in detail in the CCPS CPQRA book [Ref. 1]. A fault tree is constructed by starting with a top event or condition and asking the question:

What conditions lead to this outcome?

Each condition in the next layer down is then analyzed by asking the same question. What conditions lead to this outcome? The logic of how these intermediate conditions (termed intermediate events) are related is expressed using AND and OR logic gates. An AND gate output is true if and only if (iff) all of the inputs to the AND gate are all true simultaneously. An OR gate output is true iff one of the inputs to the OR gate is true.

Once the fault tree logic structure is created, numerical data can be used to calculate the frequency or probability of the top event of concern occurring. For fault trees where a basic event is found in multiple branches of the tree, such as power failure, the logic structure of the tree must be simplified to remove the impact of the repeated events on the top event of concern. This simplification is done using Boolean Algebra. Each resulting grouping of basic events can cause the top event to occur. These groupings of basic events are called minimal cut sets. As a simple example consider the Boolean Algebra representation of a fault tree :

$$T = BE11 + BE1 \bullet BE2 + BE1 \bullet BE3 + BE2 \bullet BE9 + BE3 \bullet BE7 \qquad \text{(Eq 32)}$$

Where:

T = Fault Tree Top Event

BE11 = Basic Event 11

BE1 = Basic Event 1

BE2 = Basic Event 2

BE3 = Basic Event 3

BE7 = Basic Event 7

BE9 = Basic Event 9

+ = Boolean Algebra Addition Symbol

• = Boolean Algebra Multiplication Symbol

For this example the top event T, the minimal cut sets are:

BE11

BE1 BE2

BE1 BE3

BE2 BE9

BE3 BE7

For example the existence of basic events BE1 and BE3 guarantees that the top event T will occur. Publically available standard computer software such as SAPHIRE [Ref. 8] may be used to find the minimal cut sets of a large fault tree.

The representation of the top event in equation 32 can now be used to compute the probability or frequency of occurrence of the top event. If the devices describe by the basic events are non-repairable and are tested after a period of time TI, the probability of top event, T, may be calculated as:

$$\text{Prob (T)} = \lambda_{11}\ TI/2 + (\lambda_1\ TI/2)(\lambda_2\ TI/2) + (\lambda_1\ TI/2)(\lambda_3\ TI/2) + (\lambda_2\ TI/2)(\lambda_9\ TI/2) + (\lambda_3\ TI/2)(\lambda_7\ TI/2) \quad \text{(Eq 33)}$$

Where:

λ_{11} = failure rate of device 11 in basic event BE11, failures per unit time

λ_1 = failure rate of device 1 in basic event BE1, failures per unit time

λ_2 = failure rate of device 2 in basic event BE5, failures per unit time

λ_3 = failure rate of device 3 in basic event BE3, failures per unit time

λ_7 = failure rate of device 7 in basic event BE7, failures per unit time

λ_9 = failure rate of device 9 in basic event BE9, failures per unit time

TI = test interval, time

Equation 33 assumes that the device failure rates are small and the product λ TI < 0.1. The other major assumption is that the values of all of the failure rates and test intervals are known. In reality the failure rate data are not known exactly with only a range or probability distribution representing the state of knowledge.

4. Example Problem

The following example is taken from the ISA technical report on fault tree analysis [Ref. 9]. This example should not be considered as a recommendation by either ISA or AIChE. The example is presented to allow for the demonstration of the methods that can be used in a fault tree study to quantify the uncertainty in the resulting calculated failure frequency or failure probability.

Consider the interlock on the intermediate storage tank T101 shown on Piping and Instrument Diagram in Figure 1. The interlock block diagram is shown in Figure 2. The interlock is intended to prevent an abnormal condition in the tank which could lead to a uncontrolled release of the tank contents. The process hazards analysis team has recommended that the interlock be designed to safety integrity level 2 (SIL 2) with a target probability of failure on demand (PFD) of no greater than 1E-2 or a risk reduction factor (RRF) of 100. Does the proposed interlock shown on Figure 1 meet the SIL 2 target? What is the uncertainty in the predicted PFD of the interlock?

A fault tree for the failure of this interlock is shown in Figure 3. A minimal cut set analysis has been completed and the resulting minimal cut sets are presented in Table 3. We can now write the Boolean Algebra equation that represents this fault tree as:

$$T = PE + TS1 \cdot TS2 + LS1 \cdot LS2 + FT1 \cdot FT2 + FT2 \cdot FT3 + FT1 \cdot FT3 + BV1 \cdot BV2 + BV1 \cdot SOL1 + BV2 \cdot SOL2 + SOL1 \cdot SOL2 + PT1 \cdot PT2 \qquad \text{(Eq 34)}$$

Assuming that the interlock is non-repairable until tested at time TI and using lamda-time (λT) approximation of the failure rate, the Boolean representation of the fault tree may now be converted to a failure probability model as:

$$\begin{aligned} \text{Prob}(T) = {} & \text{Prob}(PE) + (\lambda_{TS1}\, TI/2)(\lambda_{TS2}\, TI/2) + (\lambda_{LS1}\, TI/2)(\lambda_{LS2}\, TI/2) + (\lambda_{FT1}\, TI/2)(\lambda_{FT2}\, TI/2) + \\ & (\lambda_{FT2}\, TI/2)(\lambda_{FT3} TI/2) + (\lambda_{FT1}\, TI/2)(\lambda_{FT3}\, TI/2) + (\lambda_{BV1}\, TI/2)(\lambda_{BV2}\, TI/2) + \\ & (\lambda_{BV1}\, TI/2)(\lambda_{SOL1} TI/2) + (\lambda_{BV2}\, TI/2)(\lambda_{SOL2}\, TI/2) + \\ & (\lambda_{SOL1}\, TI/2)(\lambda_{SOL2}\, TI/2) + (\lambda_{PT1}\, TI/2)(\lambda_{PT2}\, TI/2) \end{aligned} \qquad \text{(Eq 35)}$$

Where:

Prob(PE) = failure probability of electronic logic solver, assumed to be constant at 5E-3.

λ_{TS1} = failure rate of temperature switch 1, failures per unit time

λ_{TS2} = failure rate of temperature switch 2, failures per unit time

λ_{LS1} = failure rate of level switch 1, failures per unit time

λ_{LS2} = failure rate of level switch 2, failures per unit time

λ_{FT1} = failure rate of flow transmitter 1, failures per unit time

λ_{FT2} = failure rate of flow transmitter 2, failures per unit time

λ_{FT3} = failure rate of flow transmitter 3, failures per unit time

λ_{BV1} = failure rate of block valve 1, failures per unit time

λ_{BV2} = failure rate of block valve 2, failures per unit time

λ_{SOL1} = failure rate of solenoid valve 1, failures per unit time

λ_{SOL2} = failure rate of solenoid valve 2, failures per unit time

λ_{PT1} = failure rate of pressure transmitter 1, failures per unit time

λ_{PT2} = failure rate of pressure transmitter 2, failures per unit time

TI = test interval, time

The electronic logic solver PE is assumed to have a constant failure probability of 5E-3.

Failure rate data for the other devices are present in Table 4. The mode (most likely value) of the failure rate is set equal to the point value taken from the ISA example [Ref. 9]. The triangular probability distribution (Appendix A) was assumed to represent the failure rate data for the equipment items. The upper and lower limits are based on the failure rate data of Appendix 4 of Smith [Ref. 10].

Using the device failure rates represented by the Table 4 column labeled as the mode and the minimal cut sets presented in Table 3, the ISA technical report [Ref.9, page 35] computes an interlock PFD as 7.5E-3 (RRF of 133). This would satisfy the requirement for a SIL 2 interlock {PFD of 1E-2 or a RRF of 100}. Now what is the uncertainty in the interlock PFD?

The first thing to remember is that the PFD should be calculated based on the mean value of the failure rate lambda (λ), not the mode of the failure rate. For a data set, the mean is the best single point representation of the data set. Table 5 presents the PFD of each minimal cut set using the mean of the device failure rate lambda (λ). The total of the PFD becomes 1.47E-2 (RRF of 68). This would not satisfy a SIL 2 interlock requirement.

Now we will compute the variance of the predicted probability of top event, T. To compute the variance of the probability of the top event, T, we will need to compute the sensitivity of the top event to each of the variables.

Sensitivity Using Numerical Perturbation Calculation

We will use the ISA interlock example previously analyzed by Freeman (Ref. 7). Using public data sources Freeman found that the probability of failure on demand (PFD) was 1.49E-2. To use the perturbation method for the evaluation of the sensitivities requires evaluating the impact of small changes in the uncertain input parameters. To enable these calculations to be completed numbers using 8 significant digits were used. The calculations were completed using Microsoft Excel which uses double-precision floating-point arithmetic compliant with IEEE 754 specification. Excel nominally carries 15 significant figures in calculations. We will complete the calculations with Excel displaying 5 significant digits and round the result at the end of the calculations to 4 significant digits. We will illustrate the calculations by looking at the Flow Transmitter, FT1.

a. First we recalculate the PFD of the example and find it to be 1.4594E-02. This is the base number that we will use in the perturbation calculations.
b. We now make a small change in mean failure rate of FT1. The mean failure rate is 8E-6 failures/hour. A 1% perturbation is used or an increase of 0.08E-6 failures/hour.

c. The perturbed value of the failure rate of FT1 becomes (8 + 0.08) * 1E-6 failures/hour or 8.08E-6 failures/hour.
d. The perturbed value of the failure rate of FT1 is now used to re-calculate the PFD of the interlock. The resulting re-calculation finds that the PFD is 1.4618E-02
e. The change in the PFD from the base number due to the perturbation of the failure rate of FT1 is now computed as:

   delta PFD = 1.4618E-02 - 1.4594E-02 = 2.4556E-05

f. The sensitivity of the PFD to a change in the failure rate of FT1 is now computed by dividing the change in the PFD by the change FT1 failure rate as:

   sensitivity = (2.4556E-05)/(0.08E-6 failures/hour) = 307.0 hours
   Result is rounded to 4 significant digits.

We now note that in the paper by Freeman (Ref. 7) that equation 32 presents the sensitivity for FT1 as:

$$\frac{\partial Prob(T)}{\partial \lambda \mathrm{FT1}} = 307.0 \text{ hr} \qquad \text{(Eq 32 of Ref 7)}$$

The answers are exactly the same. The sensitivity calculations for all of the devices are presented in Table 6. Note that in every case the sensitivity found using the analytical expression and the numerical estimate are the same. This is a general result which will be true for all basic event equipment failure rates found in a fault tree minimal cut set.

Once found, the sensitivities can now be used to compute the variance contribution of each particular device to the overall variance of the top event PFD in the same manner as Freeman previously presented.

The sensitivities of the top event probability are summarized in Table 6. We may now compute the variance contribution of each uncertain variable using Equations 1 – 4 in the methodology section of this paper. The relative variance contribution is expressed as a percent of the total variance of the top event probability. The top event variance is found to be 8.24E-6 (failures per hour)^2 or equivalently the top event standard deviation is 2.78E-3 failures per hour.

In this example the variance (measure of uncertainty) is dominated by the flow transmitters. The flow transmitters account for almost 2/3 of the uncertainty in the top event probability. Failures of the flow transmitters appear in 3 of the minimal cut sets. Since the uncertainty in the transmitter failure rates is large, the flow transmitters should contribute a significant amount to the variance of the fault tree top event.

We can use the variance in the top event to define the likelihood of achieving SIL 2 performance of the interlock. The calculations are done using the normal probability distribution. The normal distribution is tabulated as the standard normal distribution using a normalization factor Z.

The standard normal factor, Z, [Ref. 11] is defined as:

$$Z = \left[\frac{xi - E(x)}{\sigma}\right] \qquad \text{(Eq 36)}$$

Where:

σ = standard deviation. Note that the variance of a random variable is the square of the standard deviation of the random variable.

$E(x)$ = Expected value of the random variable x_i

For the example, the expected value of the probability of the top event of concern was calculated as

$E[Prob(T)] = 1.47E\text{-}2$ (Eq 37)

From Table 4, the standard deviation in the probability of the top event is

$\sigma_T = 2.78E\text{-}3$ (Eq 38)

The target PFD for a SIL 2 interlock must not be worse than 1E-2 (RRF of 100).

We compute the normal distribution Z factor at the SIL 2 target value as

$Z = (0.01 - 0.0147)/2.78E\text{-}3$ (Eq 39)

$Z = -1.69065$ (Eq 40)

Tabulations of standard normal distribution are presented in most statistics books such as Meyer [Ref. 11]. For this value of Z, the corresponding probability that the interlock will achieve SIL 2 performance is 4.55 percent. We can also compute the risk reduction that the proposed interlock design will achieve.

What is the interlock probability of failure on demand (PFD) that we can be 95% certain that the interlock will provide? At the 95% level, the corresponding Z factor is 1.65 [Ref 11]. We now compute the corresponding PFD of the interlock that corresponds to this Z factor.

$$Z = \left[\frac{x95\% - E[Prob(T)]}{\sigma}\right]$$ (Eq 41)

In this case:

$x_{95\%}$ = the PFD that we can be confident that the interlock will achieve

$\sigma_T = 2.78E\text{-}3$ (Eq 42)

$E[Prob(T)] = 1.47E\text{-}2$ (Eq 43)

$Z = 1.65$ (Eq 44)

Rearranging equation 40 to find $x_{95\%}$ we obtain:

$x_{95\%} = E[Prob(T)] + Z\,\sigma_T$ (Eq 45)

$x_{95\%} = 1.47E\text{-}2 + (1.65)\,(2.78E\text{-}3) = 0.0193$ (Eq 46)

or a 95% certain risk reduction factor of

$RRF = 1/PFD = 1/0.0193 = 52$ (Eq 47)

By the same logic there is 5% chance that the risk reduction is a PFD of 0.01 or an RRF of 99. In summary, this interlock is likely to perform as a mid-range SIL 1 interlock, not the SIL 2 interlock desired by process hazards analysis team.

5. Methodology for Fault Tree Uncertainty Analysis

We can now generalize the methods used in the above example to a proposed method for the evaluation of the uncertainty in a fault tree. The following step-by-step procedure is modified from that that previously published by Freeman (Ref 7) to indicate the use of numerical methods to compute the sensitivity. Step 8 shown in *italics* indicates the modification..

1. Create the fault tree using standard methods outline in the CPQRA book [Ref 1].
2. Determine minimal cut sets by hand or using standard computer software such as SAPHIRE [Ref 12].
3. Define the needed failure rate data for each basic event.
4. Define those basic events that are considered to be uncertain.
5. For the basic events that are uncertain, define the probability distribution and associated parameters needed to numerically define the probability distribution. Appendix B of this paper presents the needed description of 4 commonly used probability distributions. In many cases the triangular distribution will be selected. The minimum, maximum and mode (most likely) parameters will be needed for the triangular distribution.
6. Compute the mean of each of the uncertain variable
7. Compute the probability or frequency of the fault tree top event of interest using the mean value for each uncertain variable.
8. *Compute the sensitivity of the top event probability or frequency using the numerical methods outlined in this paper.*
9. Compute the variance of each of the uncertain variables
10. Compute the variance contribution of for each of the uncertain variables to the top event probability or frequency using equation 6.
11. Compute the total variance of the top event of interest probability or frequency by summing all of the contributions determined in Step 10.
12. Compute the variance contribution percent of each uncertain variable by dividing the variable contribution (Step 10) by the total variance of the top event of interest (Step 11).
13. Define the level of risk the project is willing to take. What chance will the project management accept for potential failure of the interlock to achieve the desired risk reduction? In the above example, I have used a 5% risk of failure or a 95% certainty that the interlock will achieve a risk reduction factor of 52. In the example, there is a very low probability that the interlock will achieve a SIL 2 PFD of 0.01 or an RRF of 100.

Alternately you can report the 90% range for the top event of interest. For the example the 90% range starts at the 5% RRF of 99 and there is a 95% certainty that that an RRF of at least 52 will be achieved. The expected risk reduction for this interlock is an RRF of 68.

6. Conclusions

A general procedure has been presented to quantify the uncertainty in calculations of failure frequencies using the fault tree methodology. The uncertainty analysis procedure is based on the application of propagation of error and variance contribution analysis techniques to the minimal cut sets created during a fault tree study. The procedure is simple and can be incorporated into standard fault tree analysis programs such as SAPHIRE [Ref 8]. This procedure is based on perturbation calculation of the frequency of the top event from changes in the uncertain variables. Incorporation of a numerical perturbation method into standard fault tree software such as SAPHIRE would be easy to implement and would provide the means for evaluation of the uncertainty in the quantified fault tree results.

For the example interlock presented in this paper, the uncertainty in the device failure rates degrades the predicted interlock SIL from meeting SIL 2 requirements to becoming a mid range SIL 1 interlock. This is a general result. Uncertainty in design parameters will reduce the likelihood that the equipment will achieve desired results. The design engineer has two choices:

- Accept the conclusion that the interlock will not perform as SIL 2
- Re-design the interlock using more reliable components to achieve the SIL 2 target

The variance contribution analysis shown in Table 7 indicates that the major contributor to the uncertainty in the interlock performance is the Flow Transmitters. Flow transmitters certified to SIL-2 or SIL-3 performance with lower failure rates could improve the calculated PFD of the interlock.

7. References

1. Center for Chemical Process Safety , *Guidelines for Chemical Process Quantitative Risk Analysis – 2rd edition*, American Institute of Chemical Engineers, New York, 2000.
2. Frank P Lees, Loss Prevention in the Process Industries – 2rd Edition, Butterworth, ISBN 0-7506-1547-8, Oxford, UK, 1996
3. US Department of Labor, Occupational Safety and Health Administration, CFR 1910.119 - Process Safety Management of Highly Hazardous Chemicals, 1910.119(e)(2)(vi).
4. Edward B. Haugen, *Probabilistic Approaches to Design*, John Wiley, New York, 1968
5. Raymond. Freeman and Angela Summers, "Evaluation of Uncertainty in Safety Integrity Level Calculations," *Process Safety Progress,* Published online in Wiley Online Library (wileyonlinelibrary.com) DOI 10.1002/prs. 11805, 2016
6. Raymond. Freeman, "General Method for Uncertainty Evaluation of Safety Integrity Level Calculations – Part 2 Analytical Methods," *Process Safety Process*, Published online in Wiley Online Library (wileyonlinelibrary.com) DOI 10.1002/prs 111915, October 2017
7. Raymond. Freeman ,"Error Propagation and Uncertainty Analysis: Application to Fault Tree Analysis", Process Safety Progress, Published Wiley Online Library (wileyonlinelibrary.com/journal/prs) DOI: https://doi.org/10.1002/prs.12080, June 29, 2019

8. US Nuclear Regulatory Commission, "Systems Analysis Programs for Hands-on Integrated Reliability Evaluations (SAPHIRE) – Version 8," NUREG/CR-7039, Vol 3, June 2011.
9. ISA, "Safety Instrumented Functions (SIF) – Safety Integrity Level (SIL) Evaluation Techniques Part 3: Determining the SIL of a SIF via Fault Tree Analysis, ISA Technical Report ISA-TR-84.00.02-2002 – Part 3, Approved 17 June 2002.
10. David J. Smith, *Reliability Maintainability and Risk – 8th Edition*, ISBN 978-0-08-096902-2, Elsevier, London, 2011
11. Paul L. Meyer, Introductory Probability and Statistical Applications, John Wiley, New York, 1972

Table 1. Comparison of Analytical Sensitivity with Numerical Approximation with Perturbation of 10% in xi for the equation: $y = F(xi) = Axi^n + B$

Order of Equation n	Equation for Analytical Sensitivity $\frac{\partial y}{\partial x} = nAxi^{n-1}$	Sensitivity with xi = 1 and A=1	Numerical Sensitivity with $\Delta xi = 0.1\ xi$	Error %
1	$\frac{\partial y}{\partial x} = 1A(xi)^0$	1	1	0
2	$\frac{\partial y}{\partial x} = 2A(xi)^1$	2	2.1	5
3	$\frac{\partial y}{\partial x} = 3A(xi)^2$	3	3.31	-10%
4	$\frac{\partial y}{\partial x} = 4A(xi)^3$	4	4.64	-16%
5	$\frac{\partial y}{\partial x} = 5A(xi)^4$	5	6.11	-22%
6	$\frac{\partial y}{\partial x} = 6A(xi)^5$	6	7.72	-29%
7	$\frac{\partial y}{\partial x} = 7A(xi)^6$	7	9.49	-36%
8	$\frac{\partial y}{\partial x} = 8A(xi)^7$	8	11.44	-43%
9	$\frac{\partial y}{\partial x} = 9A(xi)^8$	9	13.58	-51%
10	$\frac{\partial y}{\partial x} = 10A(xi)^9$	10	15.94	-59%

Error = (Analytical- Numerical)/Analytical

Table 2. Comparison of Analytical Sensitivity with Numerical Approximation with Perturbation of 1% in xi for the equation: $y = F(xi) = Axi^n + B$

Order of Equation n	Equation for Analytical Sensitivity $\frac{\partial y}{\partial x} = nAxi^{n-1}$	Sensitivity with xi = 1 and A=1	Numerical Sensitivity with $\Delta xi = 0.1\,xi$	Error %
1	$\frac{\partial y}{\partial x} = 1A(xi)^0$	1	1	0
2	$\frac{\partial y}{\partial x} = 2A(xi)^1$	2	2.01	-1%
3	$\frac{\partial y}{\partial x} = 3A(xi)^2$	3	3.03	-1%
4	$\frac{\partial y}{\partial x} = 4A(xi)^3$	4	4.06	-2%
5	$\frac{\partial y}{\partial x} = 5A(xi)^4$	5	5.10	-2%
6	$\frac{\partial y}{\partial x} = 6A(xi)^5$	6	6.15	-3%
7	$\frac{\partial y}{\partial x} = 7A(xi)^6$	7	7.21	-3%
8	$\frac{\partial y}{\partial x} = 8A(xi)^7$	8	8.29	-4%
9	$\frac{\partial y}{\partial x} = 9A(xi)^8$	9	9.37	-4%
10	$\frac{\partial y}{\partial x} = 10A(xi)^9$	10	10.46	-5%

Error = (Analytical- Numerical)/Analytical

Table 3. Basic Events in Minimal Cut Sets For Example

Cut Set No	BE 1	BE 2
1	PE	-
2	TS1	TS2
3	LS1	LS2
4*	FT1	FT2
5*	FT2	FT3
6*	FT1	FT3
7*	BV1	BV2
8*	BV1	SOL1
9*	BV2	SOL2
10*	SOL1	SOL2
11	PT1	PT2

* Minimal cut Set with same basic event as another cut set

Table 4. Lamda (λ) Failure Rate Data With Uncertainty Limits

Device	Minimum Failures per 1E6 Hours	Mode Failures per 1E6 Hours	Maximum Failures per 1E6 Hours	Mean Failures per 1E6 Hours	Lamda Variance (Failures per 1E6 Hours)^2
Flow Transmitters FT1, FT2 and FT3	1.0	2.9	20.0	8.0	18.3
Pressure Transmitters PT1 and PT2	1.0	2.3	20.0	7.8	18.8
Temperature Switch TS1 and TS2	3.0	7.6	20.0	10.2	12.9
Level Switch LS1 and LS2	2.0	4.6	20.0	8.9	15.8
Block Valves BV1 and BV2	0.2	2.3	10.0	4.2	4.4
Solenoid Valves SOL1 and SOL2	1.0	2.3	8.0	3.8	2.3

Table 5. Probability of Minimal Cut Sets Based on Mean Failure Rates

Cut Set No	BE 1	BE 2	PFD based on mean λ
1	PE	-	5.00E-03
2	TS1	TS2	2.00E-03
3	LS1	LS2	1.52E-03
4	FT1	FT2	1.23E-03
5	FT2	FT3	1.23E-03
6	FT1	FT3	1.23E-03
7	BV1	BV2	3.38E-04
8	BV1	SOL1	3.38E-04
9	BV2	SOL2	3.38E-04
10	SOL1	SOL2	2.77E-04
11	PT1	PT2	1.17E-03

Table 6. Sensitivity of Example System PFD to Basic Event Failure Rates Based on a 1% Perturbation

Basic Event Label	Device Type	Mean Failures per 1E6 Hours	Failure Perturbation of 1% of Mean Failures per 1E6 Hours	Perturbed Failure Rate Failures per 1E6 Hours	Total System PFD Using Perturbed Failure Rate of Basic Event	Change in Total System PFD	Sensitivity of PFD to Change in Failure Rate Hours
FT1	Flow Transmitter	8	0.08	8.08	1.461848E-02	2.455603E-05	307.0
FT2	Flow Transmitter	8	0.08	8.08	1.461848E-02	2.455603E-05	307.0
FT3	Flow Transmitter	8	0.08	8.08	1.461848E-02	2.455603E-05	307.0
PT1	Pressure Transmitter	7.8	0.078	7.878	1.460560E-02	1.167179E-05	149.6
PT2	Pressure Transmitter	7.8	0.078	7.878	1.460560E-02	1.167179E-05	149.6
TS1	Temperature Switch	10.2	0.102	10.302	1.461389E-02	1.995945E-05	195.7
TS2	Temperature Switch	10.2	0.102	10.302	1.461389E-02	1.995945E-05	195.7
LS1	Level Switch	8.9	0.089	8.989	1.460912E-02	1.519596E-05	170.7
LS2	Level Switch	8.9	0.089	8.989	1.460912E-02	1.519596E-05	170.7
BV1	Block Valves	4.2	0.042	4.242	1.460037E-02	6.445958E-06	153.5
BV2	Block Valves	4.2	0.042	4.242	1.460037E-02	6.445958E-06	153.5
SOL1	Solenoid Valves	3.8	0.038	3.838	1.459976E-02	5.832058E-06	153.5
SOL2	Solenoid Valves	3.8	0.038	3.838	1.459976E-02	5.832058E-06	153.5

Table 7. Variance of Top Event Probability Due to Uncertain Device Failure Rates

Device No	Device Type	Label	Sensitivity of Top Event Probability to Device Failure Rate Hour^{-1}	Variance of Device Failure Rate (Failures per 1E6 Hour)^2	Variance Contribution to Top Event Variance	Contribution to Top Event Variance %
1	Logic Solver	PE	-	0	0.00E+00	0%
2	Temperature Switch	TS1	195.7	12.9	4.94E-07	6%
3	Temperature Switch	TS2	195.7	12.9	4.94E-07	6%
4	Level Switch	LS1	170.7	15.8	4.60E-07	6%
5	Level Switch	LS2	170.7	15.8	4.60E-07	6%
6	Flow Transmitter	FT1	307	18.3	1.72E-06	21%
7	Flow Transmitter	FT2	307	18.3	1.72E-06	21%
8	Flow Transmitter	FT3	307	18.3	1.72E-06	21%
9	Block Valve	BV1	153.5	4.4	1.04E-07	1%
10	Block Valve	BV2	153.5	4.4	1.04E-07	1%
11	Pressure Transmitter	PT1	149.6	18.8	4.21E-07	5%
12	Pressure Transmitter	PT2	149.6	18.8	4.21E-07	5%
13	Solenoid valve	SOL1	153.5	2.3	5.42E-08	1%
14	Solenoid valve	SOL2	153.5	2.3	5.42E-08	1%

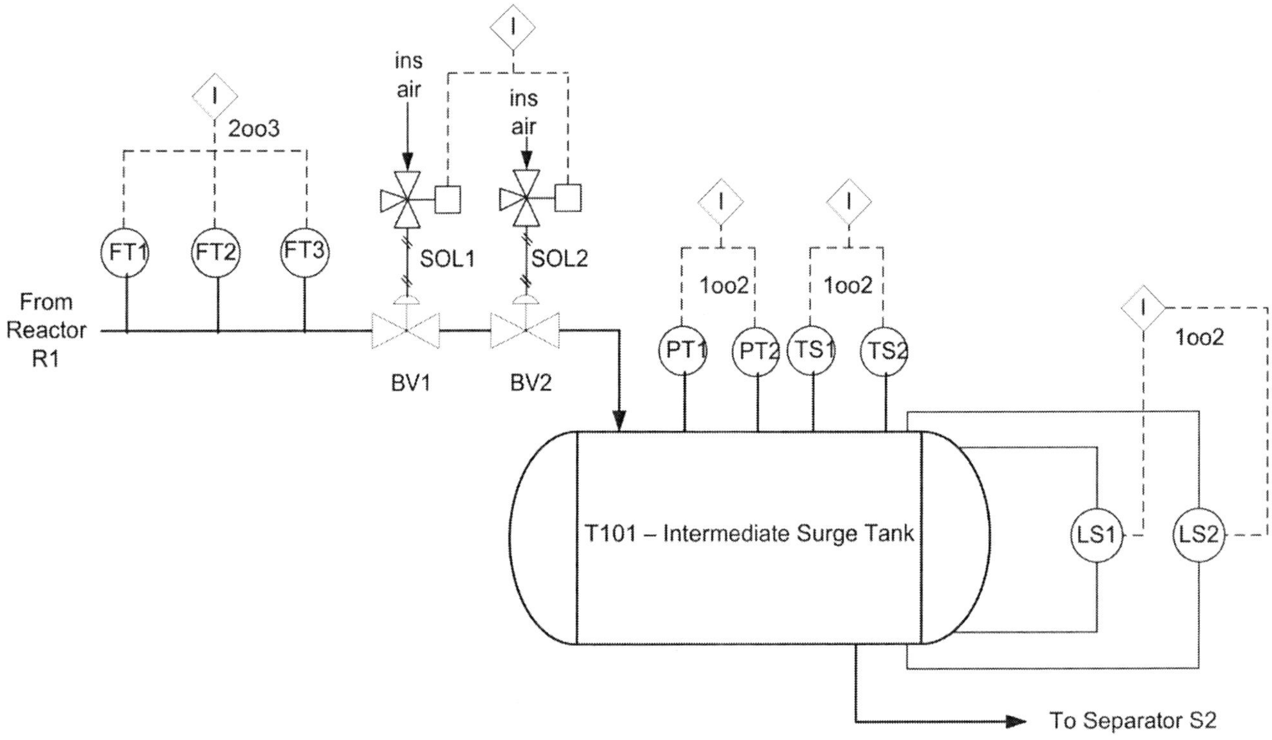

Figure 1. P&ID Diagram for Example

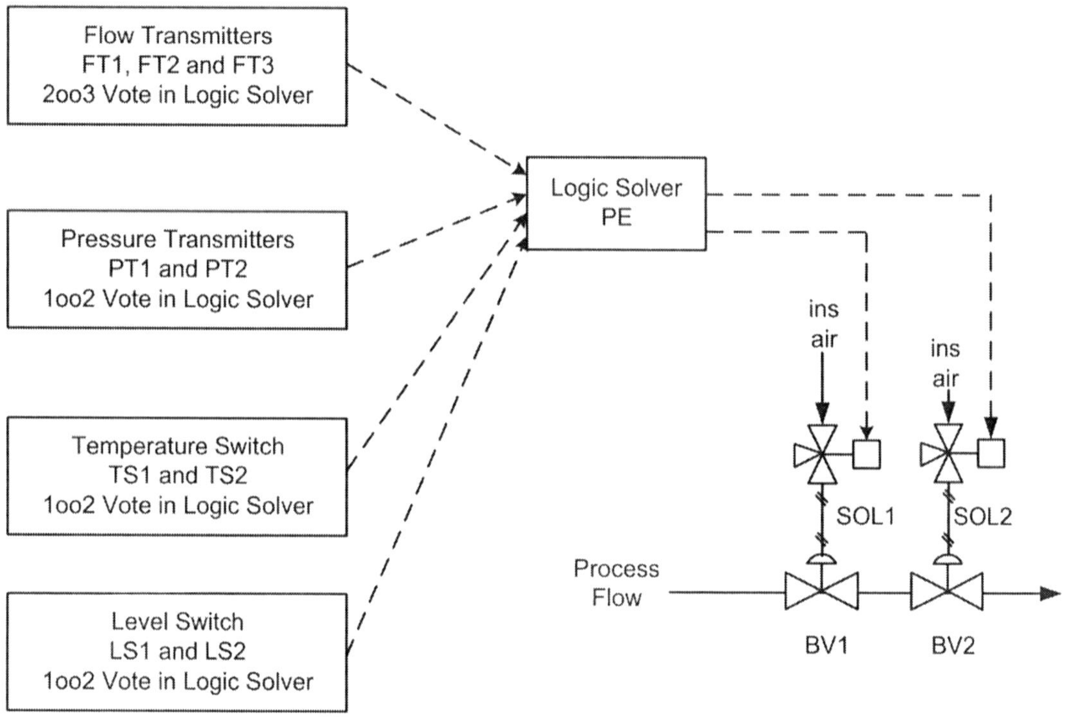

Figure 2. Interlock Block Diagram

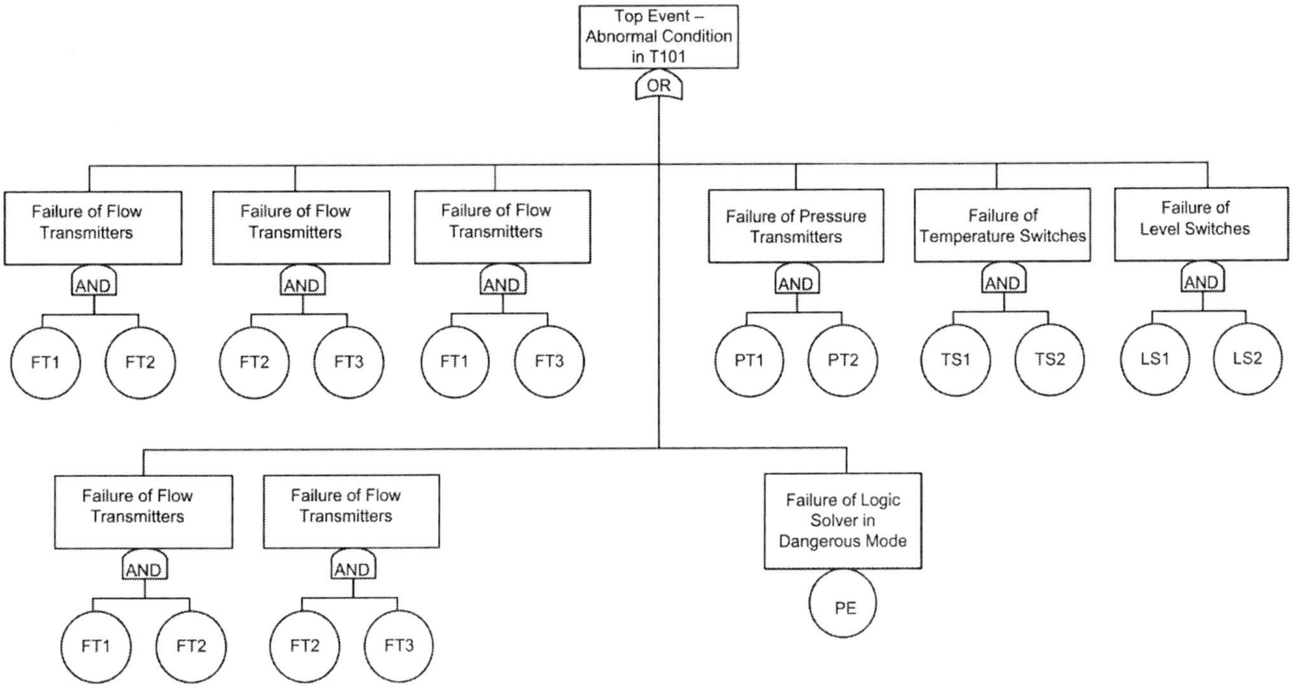

Figure 3. Fault Tree for Example Interlock

Appendix A – Commonly Encounter Probability Distributions

There are four probability distributions used to represent the failure rate of equipment items. These are:

1. Uniform Distribution
2. Triangular Distribution
3. Normal Distribution
4. Log-Normal Distribution

This appendix reviews the properties of these four probability distributions.

1. Uniform Distribution

The uniform distribution [Ref. A-1, pg 74; A-2, pg 687-688] is used when a variable is only known within a defined range. Figure A-1 presents a plot of the probability density function (pdf) for the uniform distribution. The distribution is defined using two parameters, A and B.

Where:

A = minimum value of the random variable

B = maximum value of the random variable

The mean or expected value of a random variable, x, that follows the uniform distribution is:

Mean = Expected Value – $E(x) = (A+B)/2$ (Eq A-1)

The variance of a random variable, x, that follows the uniform distribution is:

Variance = $V(x) = (B-A)^2/12$ (Eq A-2)

The standard deviation of a random variable, x, that follows the uniform distribution is:

Standard Deviation = $(\text{Variance})^{1/2} = (B-A)/(12)^{1/2}$ (Eq A-3)

2. Triangular Distribution

The triangular distribution [Ref. A-2, pg 686-687] is used when a variable is known to lie within a defined range and a "best guess" or estimate can be made as to the most likely value to the variable. Figure A-2 presents a plot of the probability density function (pdf) for the triangular distribution. The distribution is defined using three parameters, A. B and C.

Where:

A = minimum value of the random variable

B = maximum value of the random variable

C = most likely (mode) of the random variable

The mean or expected value of a random variable, x, that follows the triangular distribution is:

Mean = Expected Value – $E(x) = (A+B+C)/3$ (Eq A-4)

The variance of a random variable, x, that follows the triangular distribution is:

Variance = V(x) = $[A^2 + B^2 + C^2 - AB - AC - BC]/18$ (Eq A-5)

The standard deviation of a random variable, x, that follows the triangular distribution is:

Standard Deviation = $(\text{Variance})^{1/2} = \{[A^2 + B^2 + C^2 - AB - AC - BC]/18\}^{1/2}$ (Eq A-6)

3. Normal distribution

The normal distribution [Ref A-2, pg 665-666] is used when a variable when data analysis finds the normal distribution is the best model to describe the spread in the measured variable. Figure A-3 presents a plot of the probability density function (pdf) for the normal distribution. The distribution is defined using two parameters, μ and σ.

Where:

μ = mean of the measured data for variable

σ = standard deviation of the measured data for variable

The expected value of a random variable, x, that follows the normal distribution is:

Expected value = mean = E(x) = μ (Eq A-7)

The variance of a random variable, x, that follows the normal distribution is:

Variance = square of standard deviation = V(x) = σ^2 (Eq A-8)

The standard deviation of a random variable, x, that follows the normal distribution is:

Standard deviation = σ (Eq A-9)

4. Lognormal Distribution

The lognormal distribution [Ref. A- 2, pg 658-659] is used when a variable when data analysis finds the lognormal distribution is the best model to describe the spread in the measured variable. Figure A-4 presents a plot of the probability density function (pdf) for the lognormal distribution. Many variables in nature are lognormally distributed (for example the height of people). The log normal distribution is used when the variable of interest has a known physical lower limit of zero. A variable, x, is lognormally distributed if ln(x) is normally distributed. The lognormal distribution is defined using two parameters, μ_y and σ_y.

Let the original data be defined by the variable x with mean and standard deviation as:

μ_x = mean of the measured data for variable x

σ_x = standard deviation of the measured data for variable x

Define a new variable y as:

y = Ln(x) (Eq A-10)

Then variable x is said to be lognormally distributed if y is normally distributed. We can now write the mean and standard deviation of y (μ_y and σ_y.) [Ref. A-3, A-4 and A-5] as

$$\mu_y = Ln\left[\mu_x^2 / ((\sigma_x^2 + \mu_x^2)^{1/2})\right] \quad \text{(Eq A-11)}$$

$$\sigma_y = \left[Ln(\sigma_x^2 / \mu_x^2 + 1)\right]^{1/2} \quad \text{(Eq A-12)}$$

Where:

Ln is the natural logarithm (base e) of the argument

Appendix A References:

A-1. P. L. Meyer, *Introductory Probability and Statistical Applications*, 2nd Edition, Addison-Wesley, Reading Mass, Library of Congress Catalog no. 75-104971, 1972, pp. 74

A-2. D. Vose, *Risk Analysis–A Quantitative Guide*, 3rd edition, Wiley, Chichester, West Sussex, England, ISBN 978-0-470-51284-5, 2008, pp. 686-687

A-3. "Lognormal mean and variance - MATLAB lognstat". www.mathworks.com. Retrieved 27 December 2018

A-4. M. Mood, F. A. Graybill, and D. C. Boes. *Introduction to the Theory of Statistics.* 3rd ed., New York: McGraw-Hill, ISBN 0-07-042864-6, 1974. pp. 117.

A-5. E.L. Crow and K. Shimizu, Lognormal Distributions – Theory and Applications, Marcel Dekker, New York, ISBN 0-8247-7803-0, 1988, pp. 9

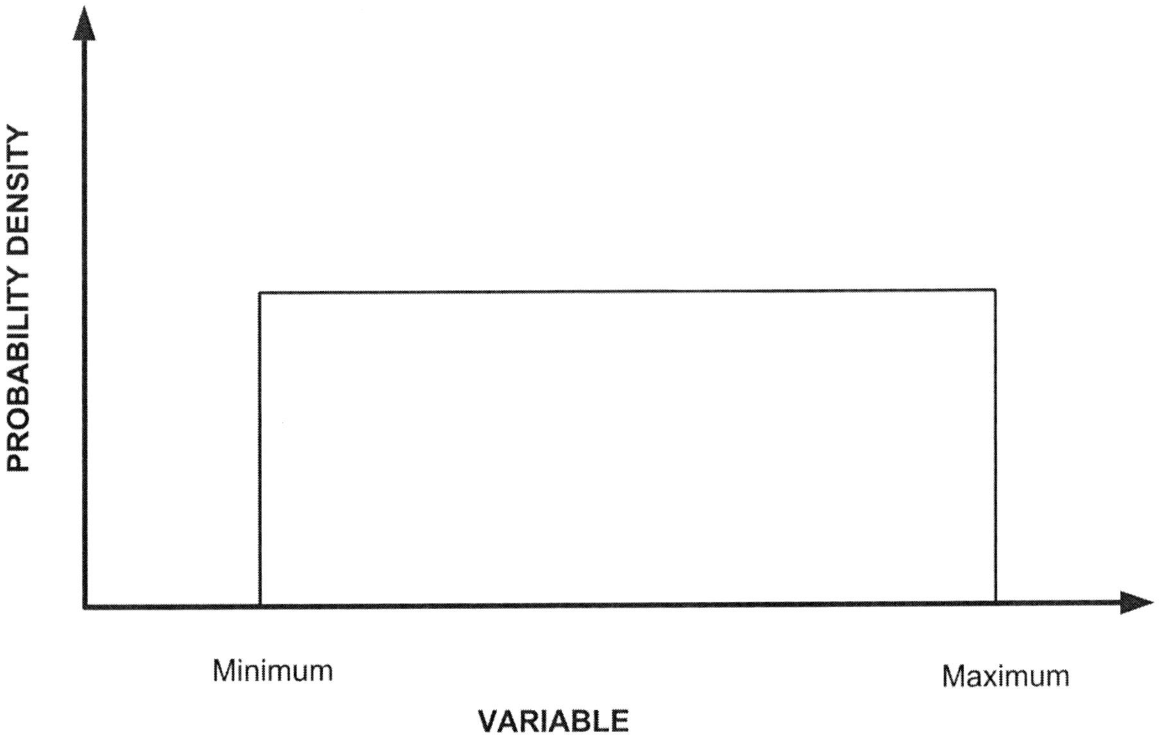

Figure A-1. Uniform Probability Distribution

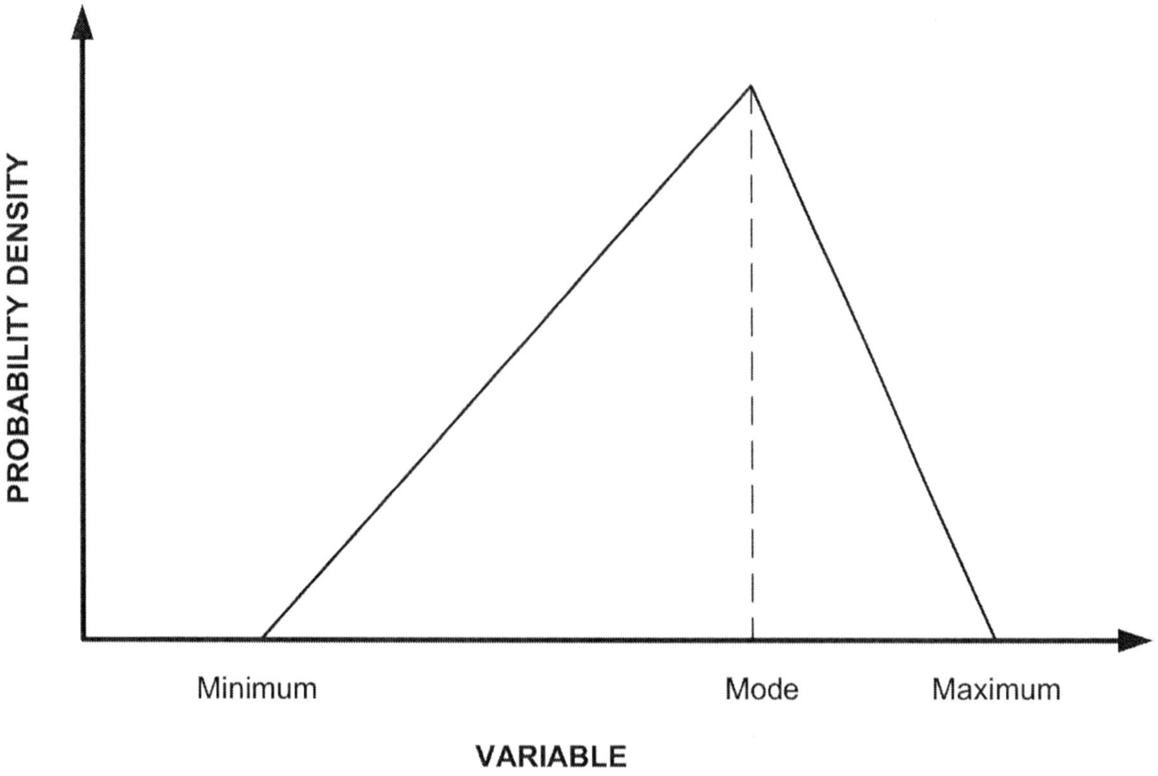

Figure A-2. Triangular Probability Distribution

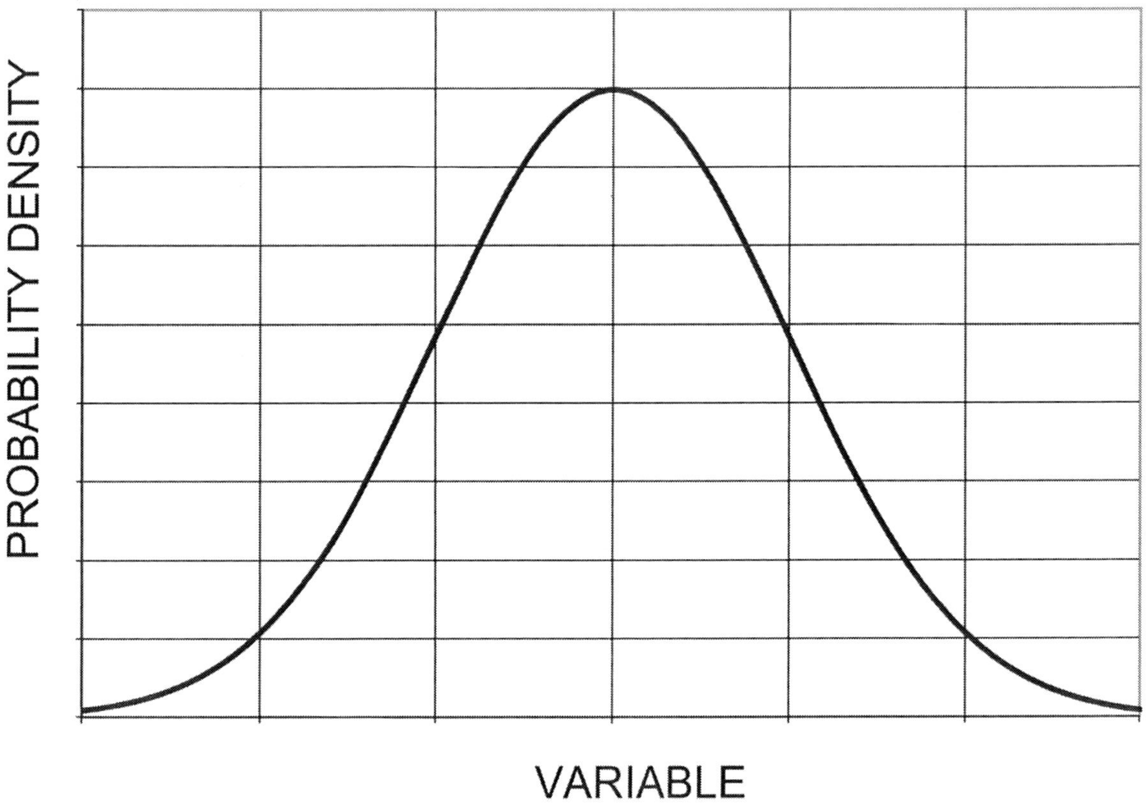

Figure A-3. Normal Probability Distribution

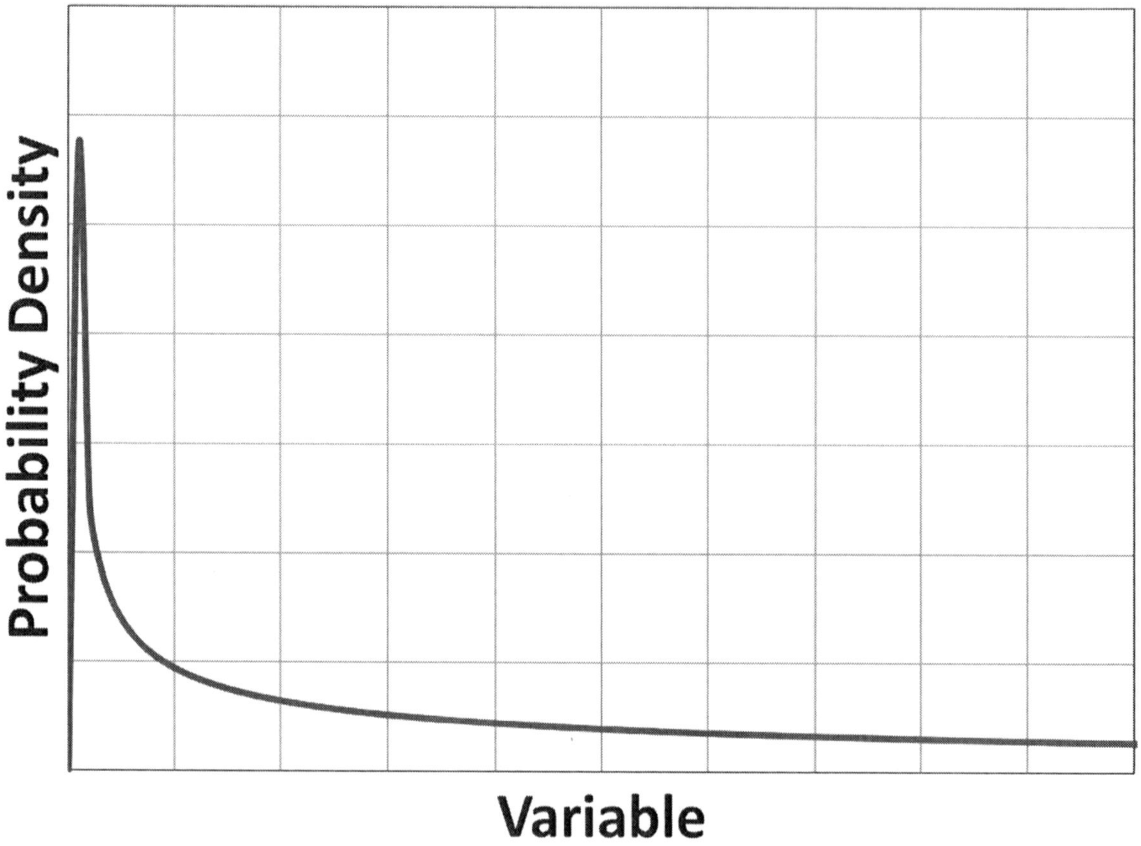

Figure A-4. Lognormal Probability Distribution

590424 Making the Most of Your PHA

Abstract:

A Process Hazard Analysis (PHA) is required to address OSHA Process Safety Management (PSM) and EPA Risk Management Program (RMP) regulations. Understanding the potential process hazard is widely recognized as a foundational element of PSM and RMP, and in addition to simply meeting regulatory requirements, can be a critical mechanism for flushing out potential vulnerabilities early in the design process for capital projects. Although an important activity, PHA is also recognized as a sizeable investment in time and energy for an Owner/Operator. So why not make the most of it? Other important activities at plants such as Mechanical Integrity, Alarm Rationalization, and Management of Change are able to stay focused and possibly be streamlined (i.e., reduced cost) based on the ability to use the results of a quality PHA having the necessary information and usability of documentation. So why not structure the PHA and documentation to be able to capitalize on potentially significant cost-savings and possible program integration?
The thrust of this paper and conference presentation will be to identify tips for conducting and documenting the PHA to maximize its ability to be used for supporting other important programs such as MOC, MI, and Alarm Rationalization. The objective will be to allow PHA Coordinators to be able to readily apply these tips to bring about improvements and cost-savings on these interrelated programs.

589814 Theoretical Analysis of the Viability of Isopropanol-Cyclohexane Solvent Systems for Biomatter Lipid Extraction

Abstract:

A theoretical alternative solvent combination for the extraction of lipids from biomatter is the Water, Isopropanol, and Cyclohexane instead of the more common combination used in the Bligh and Dyer Method (Water, Methanol, and Chloroform). While the Bligh and Dyer Method is considered effective, toxicological concerns have been raised over its use of both Chloroform and Methanol. The alternative method utilizes solvents that do not pose as high as a toxicological concern while theoretically maintaining a similar level of practical viability. The alternative method's extraction efficacy was judged against the performance of the Bligh and Dyer method using the simulation software Aspen Plus V10®. The equilibrium data generated from Aspen Plus V10® was then graphed into tertiary diagrams for data analysis using the graphing software SigmaPlot.

592166 Techno-Economic Analysis of the Conversion of CH4 and CO2 Intoacetic Acid Using Microwave Plasma

Abstract:

Methane and carbon dioxide together represent 90% of greenhouse gases [1], and one way of diminishing their emissions is through the development of pathways that convert them into added value chemicals. Acetic acid is one of such chemicals, and efforts at the laboratory scale have been carried out to obtain alternative ways of producing it. However, there are very few studies that evaluate the possibility of implementing these processes at a larger scale. In this paper, techno-economic and sustainability analyses were used to assess the feasibility of upscaling two pathways to produce acetic acid: (1) by direct conversion of methane and carbon dioxide and (2) by indirect conversion, using syngas (obtained from dry methane reforming) as an intermediate. For the direct method a microwave non-catalytic plasma reactor was considered as the main driver in the process simulation [2]. For the indirect method, a zeolite catalytical single step reaction unit was considered [3]. APEN Plus was used to simulate the reaction and purification systems of both methods, and alongside literature values and plant cost estimation techniques, the capital and operational expenses were calculated. Finally, the feasibility of both projects is assessed on carbon sequestration mass balances and economic metrics. The main sensitivity variables that need to be improved in order to obtain a minimum ROI threshold value were also evaluated.

Key words: carbon management, process design, microwave plasma reactor

[1] EPA. *Overview of Greenhouse Gases.* United States Environmental Protection Agency. 2017.

[2] Wang, L., Yi, Y., Wu, C., Guo, H., Tu, X. *One-Step Reforming of CO2 and CH4 into High-Value Liquid Chemicals and Fuels at Room Temperature by Plasma-Driven Catalysis.* Angewandte Chemie. 2017.

[3] Lopez, L., Velasco, J., Montes, V., Marinas, A., Cabrera, S., Boutonnet, M., Jaras, S. *Synthesis of Ethanol from Syngas over Rh/MCM-41 Catalyst: Effect of Water on Product Selectivity.* Catalysts. 2015.

589668 Model Predictive Control Design for the Allam Power Plant Integrated with an Air Separation Unit

Abstract:

Since electricity and heat generation account for around 25% of this greenhouse gas emission, high efficiency power generation technology such as the Allam cycle which emits essentially no CO_2 and NOx is much needed. Being an oxy-fuel power cycle, the Allam power plant is inherently integrated with an air separation plant (ASU). In this study Aspen Plus Dynamics was used to model the integrated Allam power plant/Air Separation Unit (ASU) with a high degree of heat and work integration based on the steady state work of Net Power. In our previous work, regulatory and advanced PID controls were implemented and tested for various set point tracking and disturbance rejection scenarios. However, PID controllers have limitations for various reasons. Consequently, a model predictive control (MPC) was designed in this study to optimize and improve the performance of the integrated unit and to test its performance in terms of dynamic behavior and robustness to model uncertainties.

589672 Explore the Function of Ion-Exchange Membrane (IEM) in Membrane Capacitive Deionization (mCDI) Cell, an Alternative Technique in Brackish Water Desalination Field

Abstract:

In the arid west, the freshwater supply of many communities is limited, leading to increased interest in tapping brackish water resources. Although reverse osmosis is the most common technology to upgrade saline waters, there is also interest in developing and improving alternative technologies. Here we focus on Capacitive Deionization (CDI), which has attracted broad attention as an energy efficient desalination technology, especially for ion extraction from low salinity brackish water. The objective of this work is to explore the function of ion-exchange membrane (IEM) in enhancing the performance of CDI cell. In this study, a fully coupled two-dimensional model describing ion transport and adsorption behavior in Membrane Capacitive Deionization (mCDI) cell was developed and solved numerically, taking advantage of Darcy's Law, Nernst-Planck equation, and Donnan equilibrium considering hydraulic dispersion in channel, non-ideal membrane, and modification of ion diffusivity in porous media. The model behavior was supported by the experimentally observed variations in transient effluent concentration under different operating conditions. The results showed that ion flux was significantly improved because of the IEM, leading to a faster desalination rate. The equilibrium adsorption amount is increased due to the permselectivity of the IEM, resulting in extra ion accumulation in electrode's macropores during desalination process. The cut-off time (optimal time to switch polarity to generate highest energy efficiency) of mCDI was shorter with a higher salt removal percentage compared to CDI under the same operating conditions. A sensitivity study of the thickness and the fixed charge density (starting from 500mM) of the IEM was done separately, and neither of them were found to be the dominant influencing factors under a low concentration range of brackish water (below 100mM) in this study.

590103 An Integrated Process for the Production of Propylene Glycol from Shale Gas

Abstract:

Propylene glycol (PG) is produced on a large scale primarily for the production of environmentally safest and least toxic industrial and domestic products such as polyester resins, thermoset plastics, polyurethanes, food additives and antifreeze. In this research, a novel integrated plant is developed for the production of PG from shale gas. This integrated approach has the benefit of safer operating conditions since the intermediate propylene oxide, which is explosive, does not need to be stored and transported. Furthermore, there are potential economic benefits from integration. The overall plant is simulated in the Aspenplus environment, and a variety of process conditions are tested at steady-state to optimize the production of PG. Heat-integration tools are utilized for energy-saving and capital cost reduction opportunities. A comparative economic assessment based on existing plant information indicates that the use of process integration techniques has the potential to reduce costs significantly.

Keywords: Process Integration, Heat Integration, Propylene Glycol

592456 Monitoring Methods for Fault Detection and Diagnosis: Algorithms and Applications

Abstract:

Many industrial processes collect an abundance of data from different sensors. These sensors often measure a wide variety of physical properties in order to ensure that these parameters are adhering to expected values. This is essential to ensure plant and operator safety, increase economic benefits, and maintain product quality.

A number of algorithms have been developed in order to improve existing fault detection and diagnosis performance. These algorithms integrate a number of different data-driven driven tools and methods. Multiscale wavelet-based representation of data can be used in order to handle data that is autocorrelated, non-Gaussian, and noisy. Hypothesis testing methods such as the Generalized Likelihood Ratio (GLR) technique can be used in order to provide the best possible detection for a fixed false alarm rate. Moreover, certain model-based methods have been developed in order to monitor process drifts and degradations in the process model, even when a process is operating under control.

This work will discuss the different tools utilized in development of the algorithms, their features, and practical applications, including the benchmark Tennessee Eastman Process (TEP).

589902 Selective Separation of Industrial Relevant Metal Ions from High Tds Water Using Wafer-Enhanced Electrodeionization (WE-EDI)

Abstract:

There is a need for an efficient and economical process of ion-selective separation. In hydraulic fracturing, for instance, several million gallons of water are used in a site, most of which are disposed of by deep well injection taking the water from the ecosystem and causing earthquakes. However, due to the high amount of ions in wells at fracking sites, pressure driven processes for water recovery are impossible to use. This project will demonstrate an emerging environmentally friendly Wafer Enhanced Electrodeionization (WE-EDI) technology which attempts to eliminate the shortcomings that come with normal EDI and separates effluent streams into concentrated ions and clean water. In WE-EDI, ion-exchange resins are replaced by wafers that consist of cation and anion exchange beads and a polymer binding agent. In this study, selective separation of these ion exchange wafers and the effect of electroosmotic drag on the membrane structure have been performed for various anions and cations (Na^+, K^+, Ca^{2+}, Cl^-) and improved selectivity and higher water recoveries have been observed while significant modeling has taken place to predict the selective separation as a strong function of the equilibrium constants between the ions and the beads. This project is supported by the Membrane Science and Technology (MAST) Center and is conducted with the collaboration and guidance of industry mentors.

593303 Influence of Particles on Amine Losses during CO_2 Capture: A Process Simulation Coupled Aerosol Dynamics Model _

Abstract:

Influence of Particles on Amine Losses during CO_2 Capture: A Process Simulation Coupled Aerosol Dynamics Model

David Dhanraj and Pratim Biswas

Aerosol and Air Quality Research Laboratory

Department of Energy, Environmental and Chemical Engineering

Washington University in St. Louis

St. Louis, MO 63130, USA

Efforts to reduce the concentration of atmospheric CO_2 have been initiated due to escalating concerns of global warming and climate change. Amine-based post-combustion CO_2 capture offers an economically viable retrofit for power-plants in combating these emissions. However, solvent losses have been a major challenge to commercializing this technology. Recent experiments[1] show that vapor-based and aerosol-based solvent losses are the key mechanisms of amine solvent loss during CO_2 capture. While vapor-based amine losses can be abated by employing techniques like water-washing, there has been little work to curtail aerosol-based solvent losses. This is primarily because the underlying mechanisms governing these losses are still not well understood. Although there are a few numerical models proposed for this system, there are no models that consider aerosol phenomena such as coagulation or particle loss. Moreover, these models have not been fully validated through experiments.

Furthermore, there are a few techniques that are being developed to tackle solvent losses. It is known that reducing particle number concentration in the flue gas will reduce amine emissions. Conventional Electrostatic Precipitators (ESP) are however inefficient in capturing particles in the size range 0.1 – 1.0 μm. Amongst different approaches demonstrated, introducing a pre-charging stage and a photo-ionization stage has shown potential in increasing the charging efficiency.

Therefore, the broad objectives of this work are primarily focused on two facets of the amine loss problem. The first is the development of a model that integrates aerosol phenomena with a process engineering model that accounts for the reactive mass transfer process in a packed-bed absorber. The second is the design, fabrication and testing of a bench-scale Photoionization Enhanced Two-staged Electrostatic Precipitator (PI-ESP) at Abbott Power Plant. In essence, with the PI-ESP aimed at reducing the particulate concentration in the flue gas, and the model relating the influence of particulate concentration in the flue gas to amine emissions, this work aims to illuminate the underlying mechanisms governing amine emissions, and demonstrate a solution to tackle the problem.

References

1. International Journal of Greenhouse Gas Control, vol. 85, pp. 86-99, 2019.
2. International Journal of Greenhouse Gas Control, vol. 19, pp. 138-144, 2013.

591083 Seven Questions: The Essence of Hazops

Abstract:

Before any process risk assessment can be performed, the process hazards—fires, explosions, and toxic exposures—must first be identified. This is the primary purpose of a Process Hazard Analysis, a PHA: to identify process hazards. For processes covered by OSHA's Process Safety Management (PSM) Standard, PHAs are mandated by regulation. For other processes, PHAs are a best practice.

The PSM Standard, 29 CFR 1910.119, stipulates six methods as potentially appropriate for PHAs. Some are more appropriate than other in certain circumstances; some are rarely, if ever, appropriate. The method that is most commonly used and is often the most appropriate method is the Hazard and Operability Review, the HazOp.

There are almost as many "best" ways to do HazOps as there are facilitators leading them. What they all have in common, however, is that they use worksheets that are set up as tables that look at deviations in a node. For each deviation, a HazOp team considers "Causes", "Consequences", "Safeguards", "Severity", "Likelihood", and "Recommendations", the headings of the table in a HazOp worksheet.

One of the jobs of a HazOp facilitator is to explain to team members that are new to HazOps what each heading means. Another job is to explain to experienced team members what each heading means, just in case they've picked up some bad habits.

Over the years, a set of questions has developed that are usually very helpful in helping team members understand their role in the HazOp, which is answering these seven questions.

590625 Flare Instrumentation Minimum Expectations

Abstract:

As the U.S. EPA begins expanding the flare requirements first found in the Refinery Sector Rule to other industries, newly affected facilities need to begin planning the installation of new monitoring on covered flares. This potentially includes pilot monitoring, visible emissions monitoring, vent gas flow monitoring, assist gas flow monitoring, and net heating value and/or composition monitoring; all of which will be required to meet the new requirements. This presentation will summarize the monitoring required by the new flare requirements and describe some of the technologies which may be used to comply with the regulations' requirements.

589853 Overview of Inherent Safety Assessment Tools for the Conceptual Design Stage

Abstract:

This study reviews principal concepts, tools, and metrics for risk management and Inherently Safer Design (ISD) during the conceptual stage of process design. Even though there has been a profusion of papers regarding ISD, the targeted audience has typically been safety engineers, not process engineers. Thus, the goal of this paper is to enable process engineers to use all the available design degrees of freedom to mitigate risk early enough in the design process. Mainly, this study analyzes ISD and inherent safety assessment tools (ISATs) from the perspective of inclusion in conceptual process design. The study also highlights the need to consider safety as a major component of process sustainability. In this paper, 73 ISATs were selected, and these tools were categorized into three groups: hazard-based inherent safety assessment tools (H-ISATs) for 22 tools, risk-based inherent safety assessment tools (R-ISATs) for 33 tools, and cost-optimal inherent safety assessment tools (CO-ISATs) for 18 tools. This study also introduces an integrated framework for coordinating the conventional process design workflow with safety analysis at various levels of detail.

Patentable or Infringing? An Overview of Patent Infringement and Validity Concepts for Researchers and Innovators

Michael C. Stern, Ph.D.
Exponent, Inc.
Natick, MA
mstern@exponent.com

Michael Krill, Esq.
Winston & Strawn LLP
Houston, TX
MKRILL@winston.com

Sarah Parker, Ph.D.
Exponent, Inc.
sparker@exponent.com

Abigail Oelker, Ph.D.
Exponent, Inc.
aoelker@exponent.com

Harri Kytomaa, Ph.D.
Exponent, Inc.
khytomaa@exponent.com

Prepared for Presentation at
American Institute of Chemical Engineers
2020 Spring Meeting and 16th Global Congress on Process Safety
Houston, TX
March 30 – April 1, 2020

Patentable or Infringing? An Overview of Patent Infringement and Validity Concepts for Researchers and Innovators

Michael C. Stern, Ph.D.
Exponent, Inc.
Natick, MA
mstern@exponent.com

Michael Krill, Esq.
Winston & Strawn LLP
Houston, TX
MKRILL@winston.com

Sarah Parker, Ph.D., Abigail Oelker, Ph.D., and Harri Kytomaa, Ph.D.
Exponent, Inc.

Keywords: Chemical Engineering & the Law, Legal and Process Design & Development

Abstract

This paper provides a broad overview of important concepts that researchers and innovators should be aware of when developing new techniques that may be patentable or could potentially infringe another patent. The subject matter ranges from introductory material such as defining the different parts of a patent to intermediate material including legal concepts often argued in intellectual property litigation including claim construction, infringement, and validity. Examples of these issues in the context of actual patent litigation cases relevant to chemical engineers will be provided to help connect the concepts to engineering practice.

1 What are Patents?

The U.S. Constitution gives Congress the power to enact laws relating to patents for inventions. Inventions include new and useful processes, machines, articles of manufacture, compositions of matter (e.g., Edison light bulb). The patent is a bargain between inventor and government where the inventor enhances public knowledge by disclosing how to make and use his or her invention in the patent; in return, the government grants the inventor the right to exclude others from making, using, offering to sell, or selling the invention during the life of the patent. The typical life of a patent is 20 years – and once expired, it can be freely used.

2 How do you get a Patent?

Patents are obtained by filing an application with the Patent & Trademark Office ("PTO"). A PTO examiner determines if the patent requirements are met:

- Must be **subject-matter eligible** (process, machine, manufacture, composition of matter)
  - Cannot patent abstract ideas (e.g., organizing data); laws of nature (e.g., laws of physics); or natural phenomena (e.g., cloned animal)
- Must be **useful**
- Must be **new or novel** over the "prior art"
  - Cannot be previously known, patented, or described in a publication
  - "Prior art" refers to all information available to the public before the invention that might be relevant to whether the invention is actually something new
- Must be **non-obvious** over the "prior art"
  - Even if new, cannot be obvious to try in view of the prior art

If these requirements are met, the PTO issues the patent.

3 Parts of a Patent

There are three primary parts of a patent: the cover page, specification, and claims. The cover page includes the patent number, the issue date, the inventor, the patent owner, and the filing date, the prior art and other references considered by the PTO, and an abstract of the invention.

The specification includes figures, a background of the invention, and a detailed description of the invention that may refer to the figures.

The claims are a set of numbered paragraphs at the end of the patent. These define the boundaries of a patent.

4 Patent Infringement and Invalidity

Two important concepts in patent litigation are infringement and invalidity. Infringement happens when someone makes, uses, offers, or sells a claimed invention without authorization. The accused device infringes a patent if it includes each and every element of at least one of that patent's claims. The accused device may literally include an element (i.e., exactly as described in the patent) or it may include an equivalent element. For an equivalent, non-identical element to be considered infringing, it must perform substantially the same function in substantially the same way to achieve substantially the same result as the required element.

Invalidity is a defense typically used by the party accused of infringing a patent's claims. The accused party would argue that the claims are invalid (i.e., the claims never should have been granted by the PTO), and therefore it is not liable for any damages. The basis of an invalidity argument could be that the claims are directed to an ineligible subject matter (e.g., a law of nature), that the claims are not new or novel, or that the claims are obvious. An invalidity argument can also be made if the inventor publically used or sold the invention more than one year before filing for a patent, unless the inventor did so for experimental purposes only.

5 Case Study 1

Our first case study considers the issue of what is a patentable subject matter. This relates to the patent U.S. 4,259,444 by inventor Ananda Chakrabarty, which was assigned to the General

Electric Company. The patent's claims include genetically engineering microorganisms designed to degrade elements of crude oil including camphor, octane, salicylate, and naphthalene.

Initially, the patent examiner rejected the several of the patent's claims because microorganisms were considered "products of nature" and because living things cannot be patented. The latter argument was confirmed by the PTO Board of Appeals. The appeals continued, however, until the case eventually reached the U.S. Supreme Court.[1]

The U.S. Supreme Court ruled that microorganisms were eligible for patent protection. Specifically, the opinion stated that "[a] live, human-made micro-organism is patentable" if it "constitutes a 'manufacture' or 'composition of matter.'" The Court opined that Chakrabarty's microorganism met this requirement.

6 Case Study 2

Our second case study touches on the topics of patent validity, novelty and experimentation. The story of this invention begins in 1848, when Samuel Nicolson began to experiment with wooden pavement.[2] Nicolson evaluated four different configurations of his wooden pavement system on 100 feet worth of a public road in Boston. After grading the roadbed and laying down a base layer of sand, lime and tar, Nicolson had the wooden pavement installed. The pavement system consisted of wooden blocks of various lengths and arrangements held together with nails. The spaces between the blocks were filled with crushed stone and pine tar was poured over the pavement as a preservative.[3] Nicolson's wooden pavement was in view, and in use for six years before it was patented in August of 1854.[2,4]

[1] US Supreme Court Opinions, Diamond v. Chakrabarty, 447 U.S. 303 (1980), available at: https://supreme.justia.com/cases/federal/us/447/303
[2] US Supreme Court Opinions, Elizabeth v. Pavement Company, 97 U.S. 126 (1877), available at: https://supreme.justia.com/cases/federal/us/97/126/
[3] Nicolson, S., "The Nicolson Pavement," Boston: H.W. Dutton & Son, 1859, pp. 4-7, available from the University of Michigan "Making of America" digital collection at: http://name.umdl.umich.edu/AJR3196.0001.001
[4] U.S. Letters Patent No. 11,419 at Cover

Figure 1 An example of wooden pavement.[5]

In 1878, Nicolson's company, the American Nicolson Pavement Company sued the city of Elizabeth, NJ for patent infringement. The defendants argued that Nicolson's patent was not valid because (1) it lacked novelty, and (2) the woodblock pavement of the invention was in public use before the patent was filed. This case went all the way to the U.S. Supreme Court.[2]

In its ruling on the matter, the Supreme Court argued as follows:

> "... though it may be true that every one of these elements had been employed before, in one kind of pavement or another, yet they had never been used in the same combination and put together in the same manner as Nicolson combined and arranged them, so as to make a pavement like his."[2]

The Court found that no teachings or motivation to combine the required elements to make the combination disclosed in Nicolson's patent could be found in the prior art. The Court ruled that the Nicolson patent was valid.

In addition, even though Nicolson's pavement "was publicly used for the space of six years before his application for a patent," it was clear that Nicolson "did not intend to abandon his right to a patent." The pavement in question was constructed "by way of experiment for the purpose of testing its qualities" including "durability and liability to decay." Because Nicolson had installed the pavement for experimentation instead of for earning a profit, it was not actually in public use. The Court concluded that "it is the interest of the public as well as himself that the invention should be perfect and properly tested before a patent is granted for it."[2]

There are several key takeaways from this case study. Firstly, patents are not invalid just because each individual part of a claim was known in the prior art. To invalidate a patent, you

[5] Photo credit: By Z22 - Own work, CC BY-SA 4.0, https://commons.wikimedia.org/w/index.php?curid=41758076

have to show some motivation for using the claimed combination. Further, if you start commercializing an invention or earning a profit on it, you need to file for a patent within one year. Genuine experimentation, however, does not start the one year clock.

CONTINUUM APPROACH TO OPTIMIZING DOWNSTREAM FINAL DRYING WITH UPSTREAM SOLID-LIQUID FILTRATION, CAKE WASHING & DEWATERING

Barry A. Perlmutter
President & Managing Director
BHS-Sonthofen Inc.
14300 South Lakes Drive,
Charlotte, North Carolina 28273
Tel: 1.704.814.7661
E-mail: barry.perlmutter@bhs-filtration.com
Blog: www.perlmutterunfiltered.com
Website: www.perlmutter-ideadevelopment.com

ABSTRACT

This article will review the optimization of the selection of the final drying technology in conjunction with the upstream continuous solid-liquid filtration, cake washing, and dewatering technology selection. Most often when analyzing a new process development approach, engineers take a "silo" approach and look at each step independently. This article illustrates that by taking a holistic approach and looking at each step not individually but as a continuum, the process solution becomes much more efficient.

In the manufacturing of this specialty chemical, the crystals coming from the reactor in a methanol slurry must be filtered, washed and dewatered and then dried to a final moisture of less than 1.0 (<1.0 %). The standard approach would be to first look at the solid-liquid filtration step and optimize this step for the maximum washing and drying efficiency and then with this information optimize the downstream drying. The operating company, however, took a different approach and looked at the process as a continuum from solid-liquid filtration through cake washing and dewatering to final drying. The Continuum Approach" resulted in operational energy and nitrogen savings as well as lower capital and installation costs for a more efficient and reliable process.

INTRODUCTION

The operating company had several objectives in mind for this process:

- Maximum solid-liquid filtration performance
- Low wash ratios for minimum wash media consumption
- Lowest possible residual moisture in discharged filter cake
- Final moisture of <1.0%

The BHS technologies of continuous solid-liquid separation as well as with the AVA drying technologies are in the unique position to meet the above objectives.
The paper begins with a discussion of the BHS technology and testing of the Rotary Pressure Filter. It continues with a discussion of the AVA drying testing and technology. The optimization testing is then described. Finally, the results are presented illustrating an optimum process design.

BHS ROTARY PRESSURE FILTER (RPF)

The rotary pressure filter, shown in Figure 1, is a continuous pressure filter designed for thin cake to deep cake filtration with cake depths from 6 mm to 150 mm. The slowly rotating drum (6 – 60 rph), in Figure 2, is divided into segments (called cells) each with their own filter media (synthetic cloth or single or multilayer metal) and outlet for filtrate or gas. The outlets are manifolded internally to a service/control head where each stream can be directed to a specific plant piping scheme or collection tank. In this way, the mother liquor can be kept separate from the subsequent washing filtrates and drying gases. This allows for better process control as well as reuse and recovery of solvents and the gases. The service/control head, for this application, is pressure rated so the liquefied gas can be kept under pressure, acting as a liquid.

Figure 3 shows a typical operation as follows. The feed suspension enters each cell, under constant pressure, to form a filter cake. Internal divisions of the housing allow the cake to be processed in separate zones. Each zone can operate under different pressures depending upon the compressibility of the cake at each stage of filtration, washing and drying.

For example, pressure filtration is conducted up to a maximum differential pressure of 6 bar. Pressure filtration has the added benefit of eliminating post-precipitation of the solids if the process solvent is prone to flashing under vacuum filtration. In this case, the 6 bar pressure differential allows for a back-pressure of 2-bar or more, such that filtration can be at 4 bar and the gas can be a liquid.

A second benefit of the slow-rotating pressure filtration is the reproducible cake depth and the ability to control the residence time (by the speed of the drum). The slow rotation also results in much lower maintenance and energy costs.

Positive displacement washing or counter-current washing follows filtration. The design of the pressure cells, such that they have no free-space and are completely filled, eliminates the possibility of cake cracking or bypass, providing maximum washing efficiency. Multiple washing steps can be accomplished along with solvent exchanges, steaming and extraction.
The cake then undergoes a drying step by blowing ambient temperature or hot gas through the cake. The control of this drying step is influenced by gas flow, pressure, temperature as well as the cake depth and residence time.

Finally, the pre-dried cake is continuously discharged. The cake chute is pressurized which allows the cake to be fed directly into the downstream dryer without the use of rotary locks. This is an important benefit to keep the entire system pressurized. The cake discharge is assisted by a gas pulse.

After cake discharge, there is a cloth rinse step. This is controllable depending upon the product discharge, filter media and potential of cloth blinding. With the drum rotating, after the cloth rinse step, the cells are ready to be filled for filtration. A typical process flow is shown graphically in Figure 3.

ON-SITE PROCESS TEST WORK: POCKET LEAF FILTER (PLF) TEST WORK

Process testing was conducted at the site's laboratory and in the plant. For the bench-top lab testing, the BHS pressurized pocket-leaf filter (PLF) with 20 cm^2 of filter area is used as shown in Figure 4. For the continuous pilot testing, a pilot RPF with 0.18 m^2 of filter area is installed, in Figure 5.

The objectives of the PLF testing are as follows:

- Filtration time vs. cake thickness
- Filtrate quality vs. filter media
- Cake solids wash time and quality
- Cake solids drying time and final moisture

CONCLUSION FROM THE LAB & PILOT TESTING AND SCALE-UP

The PLF tests demonstrated that acceptable filtration and solids wash rates could be obtained for this product and acceptable solids levels were observed for the mother liquor filtrate. Washing results and drying quality were also achieved.

The pilot tests confirmed the lab testing such that with a 14 micron cloth and a cake thickness of 25 mm, filtration times and filtrate quality were achieved. The wash ratios were very efficient, ranging from 0.7 to 1.2 kg Methanol/kg dry solids.

The moisture content of the cake varied between 11 – 30% depending upon the amount of nitrogen for blowing of the cake for drying. Based upon the initial thinking, the sizing of the Rotary Pressure Filter for moisture of 11% resulted in a filtration area of 2.88 m2, the RPF model, A-12 and nitrogen solvent recovery package to reduce the nitrogen usage. This is the important point which led to the optimization testing of the AVA dryer.

AVA VERTICAL BATCH DRYERS

The AVA vertical batch dryer, Figure 6, is made up of a conical vessel with a dished head and short cylindrical bottom part. The dryer consists of a conical mixing helix a centrally arranged shaft. Large radii of the welds and round bars allow for good cleanability and very close tolerances to the vessel walls. The helix is mounted in the dished head which avoids product-contacted shaft seals and eliminates possible product contamination.

Depending on the application, feed and cleaning nozzles, measuring instruments and vapor filters are also located on the vessel head. After the drying process the product is discharged via a large opening in the vessel bottom. The conical shape ensures maximum discharge of the solids - an essential aspect in the handling of bulk material.

Many solids are sensitive to mechanical stress and heat. The conical helix distributes the mechanical stress and heating uniformly in the product resulting in no hot-spots. This means that the dryer can be operated efficiently with the highest possible energy input without causing harmful stress peaks. Due to the large heating surface of the vessel jacket and the heated helix, optimum heat transfer is achieved for even and quick heating.

After the process, the product is discharged via a cleanable ball valve or knife-gate valve. To prevent cross-contamination, for example, CIP cleaning nozzles ensure that the entire interior and/or the vapor dust filter housing are operated when the machine is closed. The vertical batch dryer is well suited for the processing of other, high quality products, such as catalyst materials and metal powders.

PROCESS DRYING TESTING IN THE AVA TEST CENTER

The wet cake from the BHS Rotary Pressure Filter was tested on a vertical batch dryer in the AVA test center, Figure 7. This is where the benefits of the "continuum" approach really shine. The AVA testing started with three initial moistures at 11.6%, 17.5% and 30%. The results are as follows:

Initial moisture	Max. product temp.	Vacuum (mbar)	Drying time	Final moisture	Dryer Size
17.5%	65°C	100 - 300	45 min	<1.0%	2.4 m3
11.6%	95°C	5-120	35 min	<1.0%	1.9 m3
30 %	65°C	100 - 300	63 min	<1.0%	3 m3

The conclusion from the testing shows a small increase in the drying time and dryer sizing from a cake moisture of 30% moisture as compared with 11.6%. The final powder is in Figure 8.

CONTINUUM APPROACH & SUMMARY

With the data from BHS and AVA, the overall sizing was reviewed. Initially, with a traditional approach, the client needed 2.88 m2 of filter area plus a nitrogen-solvent recovery package for a cake of 11% moisture with the downstream dryer of 1.9 m3.

However, with the "Continuum Approach", the filter sizing can be reduced to 1.44 m2 of filter area with much less nitrogen consumption to produce a wet cake of 30% moisture. The downstream dryer increases to 3.0 m3 with only an increase in drying time to about 60 minutes which is within the client's batch time.

In summary, the "Continuum Approach" resulted in savings of energy and nitrogen plus capital savings of a smaller filter area and a small increase in the dryer capacity.

As operating companies develop new and unique chemical processes, there are many choices for filtration, cake washing, dewatering and drying. Engineers must evaluate all outcomes to make an informed and successful decision. Technical evaluation and, as shown, laboratory and pilot testing are critical for a successful decision and project. The take-away is that close collaboration between the operating company and the vendor will allow for creative problem-solving and process solutions to achieve the desired quality and production requirements.

CONTINUUM APPROACH
TO OPTIMIZING DOWNSTREAM FINAL DRYING WITH UPSTREAM SOLID-LIQUID FILTRATION, CAKE WASHING & DEWATERING

FIGURE 1: BHS ROTARY PRESSURE FILTER (RPF)

FIGURE 2: ROTARY PRESSURE FILTER DRUM AND CELLS

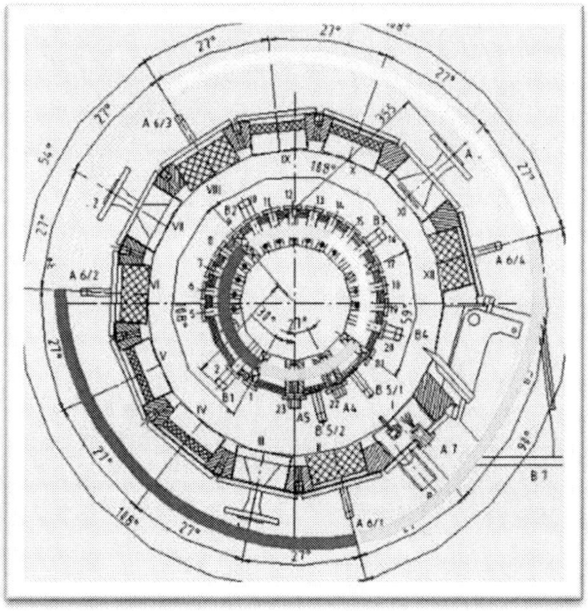

FIGURE 3: ROTARY PRESSURE FILTER PROCESS FLOW

FILTRATION (RED)
CAKE WASHING (BLUE)
DRYING (YELLOW)
CAKE DISCHARGE & CLOTH RINSE (GREEN)

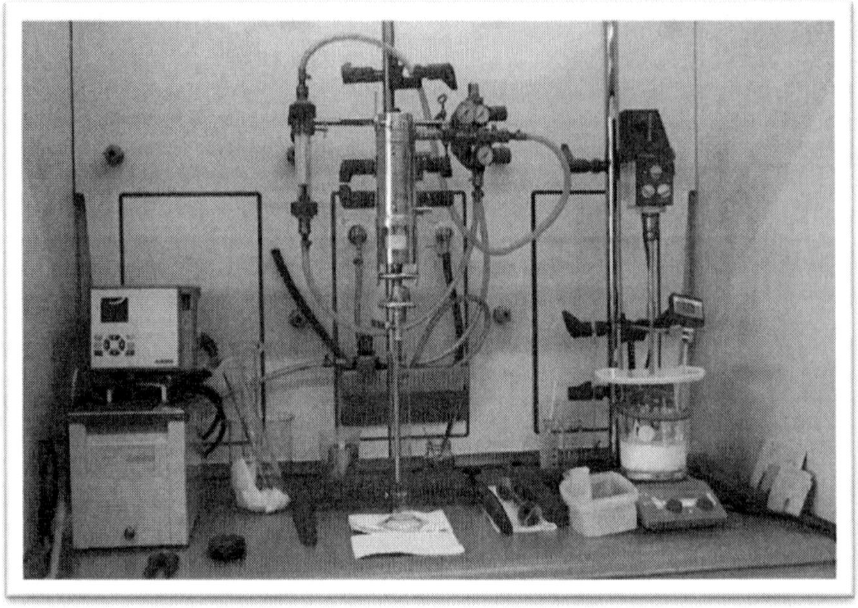

FIGURE 4: BHS POCKET LEAF FILTER (PLF)

FIGURE 5: ROTARY PRESSURE FILTER PILOT UNIT, 0.18 M^2

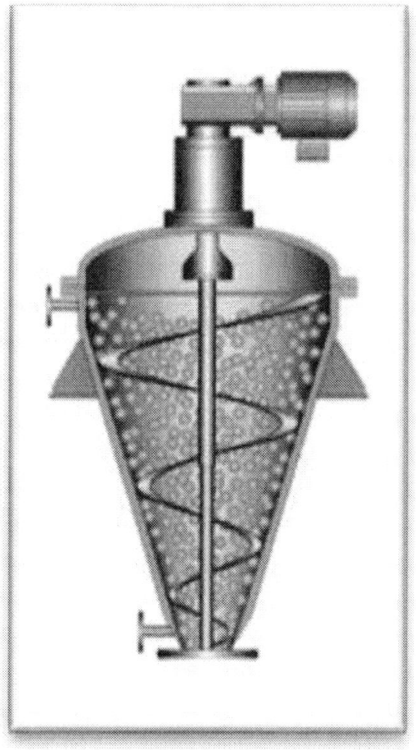

FIGURE 6: AVA VERTICAL HELIX DRYER

FIGURE 7: AVA VERTICAL HELIX DRYER IN TEST CENTER

FIGURE 8: FINAL POWDER FROM AVA VERTICAL HELIX DRYER

CONTINUUM APPROACH
TO OPTIMIZING DOWNSTREAM FINAL DRYING WITH UPSTREAM SOLID-LIQUID FILTRATION, CAKE WASHING & DEWATERING

589600 Design of Milli-Scale Flow Reactor Equipped with Additive Manufactured Static Mixer

Abstract:

Micro-reactor is an enabling technology that fulfills the requirements of small scale continuous production. Conventional static mixers cannot provide sufficient adequate mixing, mass transfer, and heat transfer at flow rates in the order of mL/min. Although there are various types of micro-reactor, and some of these reactors are on the millimeter scale, not all reactions require such degree of enhancement of mass transfer and heat transfer. Due to the fixed structure, the enhancement degree of mass transfer can only be adjusted by flow rate. All of the above factors restrict the application of flow chemistry from laboratory to production.

A series of milli-scale flow reactors equipped with static mixers manufactured by 3D printing was developed. The mixing performance of the reactors was evaluated in a wide flux range(40ml/min to 600ml/min). Mixing times in the reactors determined by Villermaux–Dushman method range from 0.03 s to 0.002 s. Flow fields in the reactors were evaluated with computational fluid dynamics and validated by particle image velocimetry. Proper milli-scale flow reactors can be selected to meet the demands of process intensification and provide a way for scale-up of flow chemistry.

590696 Bioconversion of Crude Glycerol from Biodiesel Process into Ethanol By Escherichia coli Amplified Using the PCR Techniques

Abstract:

There is large amount of unused glycerin in the biodiesel market. Biodiesel production outputs 10% w/w crude glycerol for 100% biodiesel produced. Our NSF sponsored Innovation Corps (I-Corps®) market discovery found the biodiesel industry pain points is no use for the glycerin. In order to have sustainable biodiesel plants on-site glycerin to ethanol process is needed. Having glycerin to ethanol intensifies the biodiesel process. The ethanol produced from glycerol can be used in the same plant. This will reduce logistics, and this could be used instead of methanol.

There are different ways to make ethanol but no literature that uses PCR technique to amplify the bacteria is reported. Using the polymeric chain reaction (PCR) methods the E.Coli is amplified. E.Coli strain is genetically modified so the strain that can metabolize ethanol is selected. PCR technique is easier than the conventional techniques. The resulting glycerol to ethanol rate can have an important effect on how next generation biodiesel plants will operate. Glycerol to ethanol via metabolic pathway engineering of E. Coli.

Keywords: *Escherichia coli*, biofuels, Glycerol

589924 Thermodynamic Mechanism of Self-Heat Recuperative Distillation

Abstract:

In the last several years the authors have been developing the self-heat recuperation technology for the heating and cooling thermal processes in which the heat of an effluent stream is recuperated and recycled for feed stream heating by the adiabatic compression of gas and/or vapor. Applying this technology to the distillation system, a drastic energy saving of more than 80% can be achieved.

In this study, the thermodynamic mechanism of self-heat recuperative distillation system is clarified in terms of exergy analysis by using module expression process flowsheet and temperature-entropy diagram.

589532 Estimating the Boiling Point Elevation of Kraft Black Liquor Solutions

Abstract:

Black liquor solutions from the kraft paper pulping process are comprised of an intricate mixture of both organic and inorganic salts, have a complex chemistry, and are highly corrosive due to their alkaline nature. Scaling is a significant issue during the evaporation process of weak black liquor due to the inverse temperature behavior of sodium salts. Because of the complex nature of these solutions, a rigorous thermodynamic model that can be reliably used to predict the boiling point elevation and scaling behavior is often difficult to obtain. This work seeks to correctly identify the appropriate chemistry and to use the thermodynamic framework of the electrolyte NRTL (eNRTL) model to correlate vapor-liquid and solid-liquid equilibrium data to accurately predict the boiling point elevation of black liquor solutions. This model connects the nonideality of the solutions with the composition by using two binary interaction parameters for each electrolyte-electrolyte and molecule-electrolyte pair in the system. Completion of an accurate model can aid in the mitigation of scaling issues that arise during the evaporation process and can offer a rigorous support of enhancements to the current chemical recovery cycle.

590184 Catalytic Co-Pyrolysis of Biomass and Low-Density Polyethylene with Spent FCC Catalyst

Abstract:

Catalytic co-pyrolysis of biomass and Low-Density polyethylene with spent FCC catalyst

Kanduri Praveen Kumar [a], Seethamraju Srinivas [a1]

[a]Department of Energy Science and Engineering, Indian Institute of Technology Bombay, Powai, Mumbai 400076, India.

([a] Corresponding Author: Department of Energy Science and Engineering, Indian Institute of Technology Bombay, Powai, Mumbai 400076, India; praveen.k@iitb.ac.in_)

Keywords: Biomass, Plastics, Spent FCC catalyst, Co-pyrolysis, Bio-oil

1. Introduction

Concerns over the dwindling fossil fuel reserves, rising global energy demand, and the adverse impact of carbon emissions on the global climate provide an impetus to utilize biomass as a renewable source of energy [1,2]. Biomass (B) conversion technologies such as pyrolysis and gasification are some of the pathways to convert biomass to higher-value products. Of these, pyrolysis offers an attractive way of transforming the low-energy density biomass into a high-energy-density liquid called bio-oil. Owing to its heterogeneity, thermal breakdown of biomass produces chemically different compounds and bio-oil or pyrolysis-oil has fuel characteristics that are inferior to that of conventional fuels, mainly due to the presence of oxygenated-compounds in it. Bio-oil typically contains ~45-50 wt% oxygen and is distributed widely among more than 300 compounds [3]. These oxygenates impart-high acidity, low heating value, high viscosity, and instability to bio-oil [4]. Therefore, the removal of these reactive oxygenates and improving the bio-oil quality is essential for its acceptance as a fuel.

Catalytic cracking is one of the many upgrading techniques to achieve this goal. However, the major drawbacks associated with the process are high char formation, low hydrocarbon yield, and short catalyst lifetime [5]. Although catalyst helps to reduce the oxygen content in bio-oil through mechanisms of dehydration, decarbonylation and decarboxylation, the char produced by pyrolysis reactions rapidly deactivates it. The char formation is mostly the consequence of low hydrogen-environment during pyrolysis processes, as biomass is a carbon-rich and hydrogen-deficient species. Plastics like LDPE (Low-density polyethylene) are hydrogen-rich materials [(H/C) ratio of ~2] and their utilization in pyrolysis as a co-reactant significantly improves the hydrocarbon yield, and reduces the extent of char formation. The current research focuses on the conversion of biomass to compounds of high value in the petrochemical industry. Olefins and aromatics, including benzene, toluene, and xylene are the most desired compounds in the petrochemical industry – these are the starting materials for manufacturing a wide range of specialty chemicals.

Spent FCC is a waste from petroleum refineries and is generally disposed of in landfills. Though considered as a waste, it can be used in co-pyrolysis processes to convert the biomass volatiles to more valuable compounds particularly, aromatics. This study therefore investigates the high-value utilization of low-value materials (waste biomass, waste plastics, and spent FCC).

2. Methodology

In this study, catalytic co-pyrolysis (CCP) of mixtures of biomass (groundnut shell) and plastics (LDPE) with spent FCC were performed (see Fig.1) to investigate the potential of spent FCC as a catalyst to produce aromatic hydrocarbons and to study the effect of feed composition on the pyrolysis products. The experiments were conducted at 510°C in both *in-situ* and *ex-situ* modes with 30 g of feed, 10% w/w catalyst loading, and at three different biomass to plastic ratios (1:1, 1:2, and 2:1). The pyrolysis-oil was analyzed using a GC-MS and the pyrolysis gases were analyzed offline using a Gas chromatograph equipped with a thermal conductivity detector. The results were compared in terms of % carbon yield of pyrolysis products (aromatics, olefins, alkanes, and oxygenates).

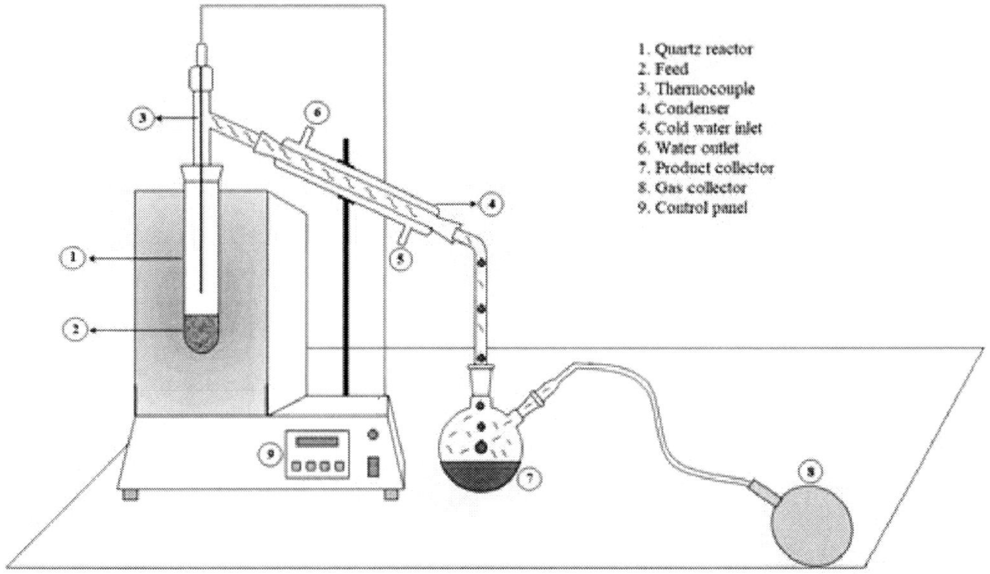

Fig.1. Schematic of catalytic co-pyrolysis experimental set-up

3. Results and Discussion

Fig. 2 summarizes the product distribution for *in-situ* CCP of biomass and LDPE as a function of feed ratio; similar plots were obtained for *ex-situ* experiments too. In both the modes, the liquid yield was maximum at higher LDPE (plastic) fractions in the feed – ~36 wt% for *ex-situ*, and ~34 wt% for *in-situ* mode. However, the aromatic yield in *ex-situ* mode was relatively higher than that of *in-situ* mode in all cases for the same feed ratios. While the aromatic yields for *ex-situ* mode were in the range of ~25-41 %C, and the peak value was observed at B/LDPE ratio of 1, the corresponding values for the *in-situ* mode were ~12-14 %C, and B/LDPE ratio of 2, respectively (see Fig. 2). The *ex-situ* mode of CCP was more efficient than the *in-situ* mode in minimizing the oxygenate content in the pyrolysis liquid. The oxygenate fraction increased with an increase in the biomass portion in the feed. The yield of oxygenates for *ex-situ* mode varied from ~14-18 %C, with a minimum value at the B/LDPE ratio of 0.5, - the corresponding values for the *in-situ* mode are ~26-30 %C at B/LDPE ratio of 1.

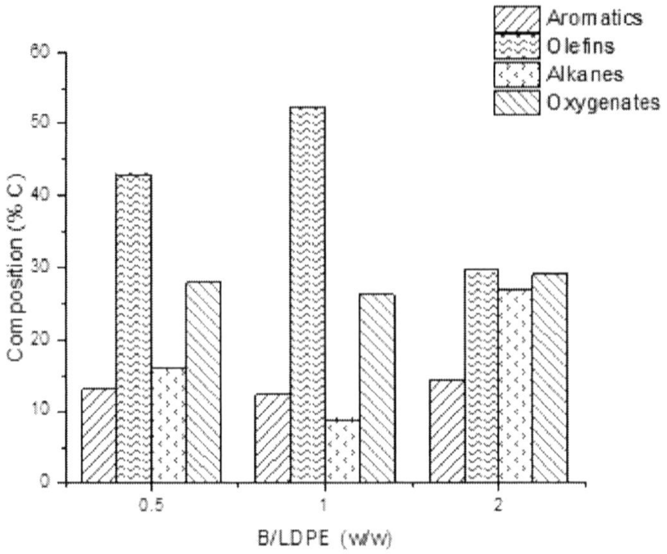

Fig. 2: Bio-oil composition as a function of feed ratio for *in-situ* CCP of biomass and LDPE mixtures

These results suggest that the product distribution in CCP is a strong function of pyrolysis configuration (*in-situ* or *ex-situ*) and feed composition. The *ex-situ* mode CCP enhanced the quality of pyrolysis liquid, especially the mono-aromatics selectivity, when compared to the liquid from *in-situ* CCP. The significant advantage of using *ex-situ* configuration is the ease of catalyst recovery which is laborious in case of the *in-situ* method. The spent FCC catalyst in both the configurations was effective in deoxygenating the biomass through dehydration, decarbonylation, and decarboxylation mechanisms. Therefore, the secondary use of this catalyst in pyrolysis will help in deriving the energy from waste biomass and plastics and also reduce the costs involved in the development and purchase of fresh or proprietary catalysts. Future work is in progress to study the effect of HDPE in CCP owing to its difference in crystallinity and thermal degradation compared to LDPE.

References

1. Dinesh Mohan, Charles U. Pittman, Jr., Philip H. Steele, Pyrolysis of Wood/Biomass for Bio-oil: A Critical Review, Energy Fuel, 20 (2006), pp. 848-889
2. George W. Huber, Sara Iborra, Avelino Corma, Synthesis of Transportation Fuels from Biomass: Chemistry, Catalysts, and Engineering, Chem. Rev., 106 (2006), pp. 4044-4098
3. A.V. Bridgwater, Review of fast pyrolysis of biomass and product upgrading, Biomass Bioenergy, 38 (2012), pp. 68-94
4. David Martin Alonso, Jesse Q. Bond, James A. Dumesic, Catalytic conversion of biomass to

biofuels, Green Chem., 12 (2010), pp. 1493–1513

5. Xiangyu Li, Jian Li, Guoqiang Zhou, Yu Feng, Yujue Wang, Gang Yu, Shubo Deng, Jun Huang, Bin Wang, Enhancing the production of renewable petrochemicals by co-feeding of biomass with plastics in catalytic fast pyrolysis with ZSM-5 zeolites, Appl. Catal., A, 481 (2014), pp. 173–182

A Corresponding States Approach to Liquid Heat Capacity QSPRs

Joey Bloxham
joeybloxham@byu.edu

BYU

Map

1. Define QSPRs
2. Discuss regression training and testing
3. Physical interpretations of correlated descriptors
4. Suggest a new equation to fit liquid heat capacity from corresponding states methods
5. Discuss success of methods

BYU

Motivations

- DIPPR 801
- Research group focus is physical properties (prediction, measurement, and understanding)
- This work was undertaken to get a qualitative understanding of liquid heat capacity (LCP), but quickly discovered predictive capabilities

BYU

Uses in Process Design

- Liquid heat capacity is an important design consideration
- Needed for design of heat exchangers, reactors, etc.
- Accurate prediction of LCP allows better process design

BYU

Defining a QSPR

- Quantitative-Structure-Property-Relationships (QSPRs, sometimes QSARs) use regression techniques to relate geometric, topological, functional group counts, and other descriptors to chemical properties
- Always expressed as single value
- Thousands of descriptors
- QSPRs accurate predictions of many properties (boiling point, viscosity, solubility, etc.)

Van der waals volume v. LCP

- Naef (2019) show linear correlation between molecular volume and LCP at 298.15 K
    - Also note this trend is possible at other temperature ranges
- Molecular volume hard to define, other descriptors needed

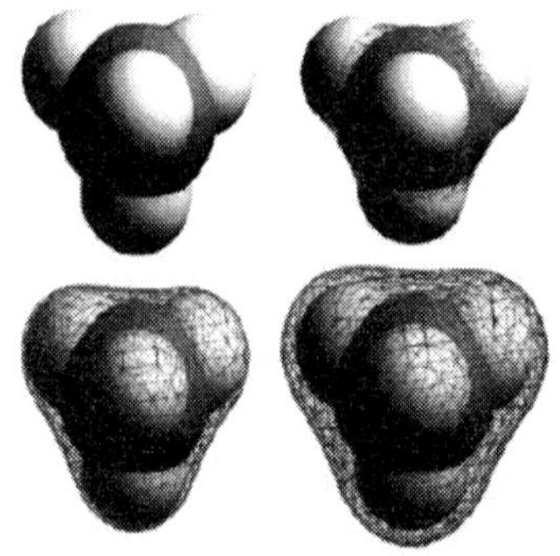

Methane molecular surface with 4 different electric cloud densities. Relick (1997)

QSPR Design

- DIPPR database has property data for 2730 chemicals
- Database was searched for chemicals with experimental data collected at 200, 300, and 400 K
  - If experimental data within 25 K of target T, correlation value at target T was used
  - Total studied species
    - 165 at 200 K
    - 639 at 300 K
    - 876 at 400 K
  - Species come from many chemical families, organic and inorganic

BYU

QSPR Design

- 5270 descriptors were calculated using Dragon 7
  - Optimized chemical structures used, from *Gaussian 09* using B3LYP/6-11+G(3df,2p)
- Species divided into training (80%) and testing (20%) groups
- Species at or above .8 T_r were eliminated

BYU

Model Choice

- Each descriptor then regressed to several different models
  - Various hyperbolic, logarithmic, polynomial, and exponential models were used, at least 15 unique models attempted
  - Both sum squared error and sum of absolute error objective functions tested
- Linear models were found to give best average absolute error and standard deviation in error
- Single-parameter linear models chosen
  - At all temperatures, meaningful descriptors had high collinearity (VIF well over 10)

BYU

Parameter Optimization

- Parameters were found 15 times from random testing/training arrangements, then averaged
- Accuracy values are using optimized parameters, averaged over 15 random arrangements
  - This method used to avoid using a "preferred fit" for training and testing data

BYU

Significant Descriptors – General Trend

- Most highly correlated descriptors tended to describe:
  - Size
  - Atomic electronegativities
  - Molecular volume
  - Ionization potentials
- These trends make sense, as there are known trends between LCP and molecular size and intermolecular forces

Data on 200 K

Training Set	%
Average Error	0.03
Average Absolute Error	10.38
Testing Set	**%**
Average Error	-0.50
Average Absolute Error	9.97

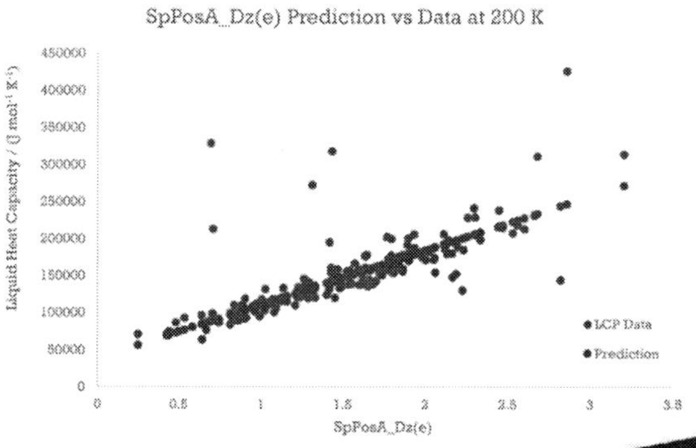

Data on 300 K

Training Set	%
Average Error	0.78
Average Absolute Error	8.18
Testing Set	%
Average Error	0.46
Average Absolute Error	8.43

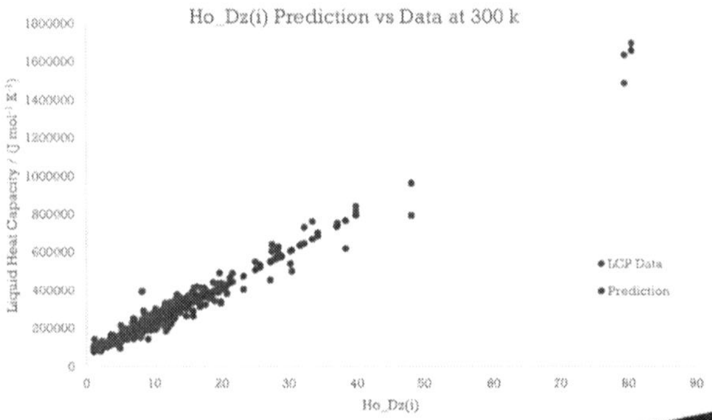

BYU

Data on 400 K

Training Set	%
Average Error	0.10
Average Absolute Error	9.43
Testing Set	%
Average Error	0.23
Average Absolute Error	9.77

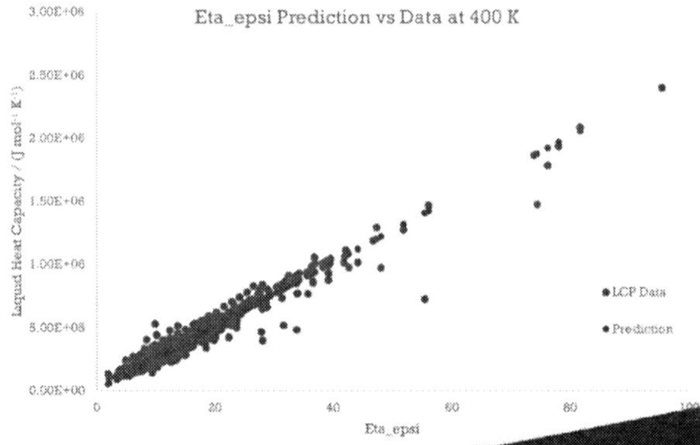

BYU

Problems with Other LCP Predictions

- Many are group contribution methods
  - This creates problems if groups are unavailable

BYU

Benefit of QSPR - atypical molecules

- Group contribution methods often skip inorganics, salts, etc.
- Descriptors available for these
- Siloxanes cannot be used on many group contribution methods
  - With Eta_Epsi QSPR at 400 K, silicon-containing compounds have average absolute error of 12.63%, and average error of -0.76%

BYU

Problems with Other LCP Predictions

- Often, methods only give LCP at single T
- If there is T-dependent portion, often ignores critical temperature effects
  - Popular methods like Ruzicka-Domalski use 2nd order polynomial fit
  - This leads to poor extrapolation at high reduced-T

BYU

Can QSPR Overcome T-Dependence?

- This work gives QSPRs with good performance at several temperatures, and capability to quickly create QSPRs at any temperature with sufficient data
- How to extrapolate between different QSPRs?

BYU

Potential Solution

- Use QSPRs to inform a LCP model inspired by corresponding states
- Proposed model – modified Lee-Kesler residual heat capacity expression
- $C_{pL} = A + \frac{B}{(1-T_r)} + C\frac{(1-T_r)^{\frac{1}{3}}}{T_r}$
  - (Since constant pressure CP used in regression, equation gives constant P LCP)
- Fitting this equation to QSPRs theoretically gives appropriate behavior near critical temperature
  - Also provides flexibility to use other prediction methods for well-suited species

BYU

Theoretical Basis of Eqn

- Rowlinson residual heat capacity function is:
- $\frac{C_{pL}-C_p^{\circ}}{R} = 1.45 + \frac{.45}{(1-T_r)} + .25\omega\left[17.11 + \frac{25.2(1-T_r)^{\frac{1}{3}}}{T_r} + \frac{1.742}{(1-T_r)}\right]$
- Treating the constants as variables:
- $\frac{C_{pL}-C_p^{\circ}}{R} = a + \frac{b}{(1-T_r)} + c\omega\left[d + \frac{e(1-T_r)^{\frac{1}{3}}}{T_r} + \frac{f}{(1-T_r)}\right]$
- And combine like terms
- $\frac{C_{pL}-C_p^{\circ}}{R} = (a + c\omega d) + \frac{(b+c\omega f)}{(1-T_r)} + \frac{c\omega e(1-T_r)^{\frac{1}{3}}}{T_r}$

BYU

Theoretical Basis of Eqn

- $\frac{C_{pL}-C_p^{\circ}}{R} = (a + c\omega d) + \frac{(b+c\omega f)}{(1-T_r)} + \frac{c\omega e(1-T_r)^{\frac{1}{3}}}{T_r}$
- Rearranging
- $C_{pL} = C_p^{\circ} + R(a + c\omega d) + \frac{R(b+c\omega f)}{(1-T_r)} + \frac{Rc\omega e(1-T_r)^{\frac{1}{3}}}{T_r}$
- Let $\mathrm{A} = C_p^{\circ} + R(a + c\omega d)$, $\mathrm{B} = \mathrm{R}(b + c\omega f)$, and $\mathrm{C} = Rc\omega e$
- $C_{pL} = A + \frac{B}{1-T_r} + C\,\frac{(1-T_r)^{\frac{1}{3}}}{T_r}$

BYU

Ideal Gas Assumption

- Ideal gas heat capacity (C_p°) was treated as constant
- In temperature range studied (200-400 K) most species C_p° well represented by first-order linear model
- Early tests used $C_{pL} = A + \frac{B}{1-T_r} + C\,\frac{(1-T_r)^{\frac{1}{3}}}{T_r} + DT_r$ to account for C_p°, but no improvement in fit was found
- Equation has enough flexibility to fit changes in C_p° in B and C terms

BYU

Example

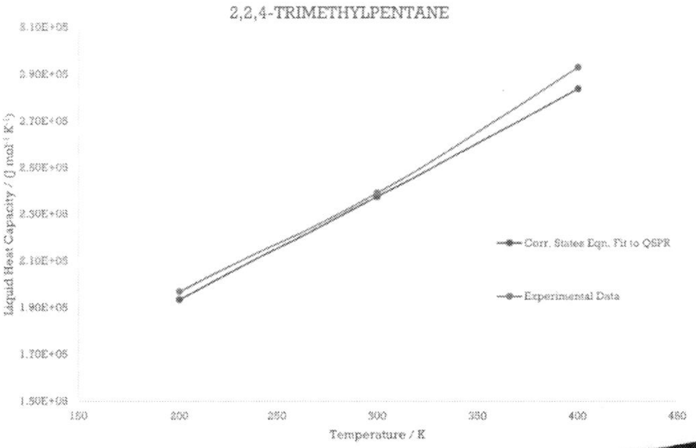

BYU

Performance at $.9\ T_r$

- Need to test performance at high reduced temperatures
- Unfortunately, not enough experimental LCP data at .9 Tc to get estimates on accuracy

BYU

Future Work

- Test high reduced-temperature fit
- Additional fitting constraints for high reduced temperature
- Splitting data sets into number of oxygens increases accuracy – easy way to improve method

Acknowledgements

- Daniel Hill
  - PhD Student, PRISM Group, BYU
- Dr. W. Vincent Wilding
  - Associate Dean of Engineering and Technology, BYU
- Dr. Knotts
  - Professor of Chemical Engineering, BYU
- Dr. Giles
  - DIPPR 801 Project Coordinator

Bibliography

- Naef, R. Calculation of the Isobaric Heat Capacities of the Liquid and Solid Phase of Organic Compounds at and around 298.15 K Based on Their "True" Molecular Volume. *Molecules* **2019**, *24*, 1626.
- Rellick, L. M.; Becktel, W. J. Comparison of van der Waals and semiempirical calculations of the molecular volumes of small molecules and proteins. *Biopolymers* **1997**, *42*, 191-202.
- Rowlinson, J. S. Liquids and Liquid Mixtures, 2nd ed., Butterworth, London, **1969**.

BYU

ABSTRACT

As part of our ongoing work to improve thermodynamic consistency in property prediction for associating fluids, an investigation into molecular descriptor-based prediction methods was undertaken. Quantitative Structure-Property Relationships (QSPRs) have found applications in many areas of chemistry and engineering as effective prediction methods. QSPRs use molecular descriptors to simplify complex molecular properties to a single value. These values can then be related through linear or non-linear models to physical properties. Melting point, viscosity, solubility, and other properties have been well represented using molecular descriptors in QSPRs.

Liquid heat capacity (LCP) is another property where QSPRs can be helpful prediction tools. Researchers have shown strong correlation between the LCP and various molecular descriptors, but these predictions are limited to a single temperature, usually 298.15 K. Engineers often need LCP prediction methods that can be extrapolated through a large temperature range for safe process design, and currently available QSPR methods for LCP lack this ability. Another important consideration is the steep increase in heat capacity as the liquid nears its critical temperature, an issue that is not accounted for in traditional polynomial-based predictions, such as the Ruzicka-Domalski group contribution method.

In this work, multiple QSPRs at various temperatures are combined using a modified corresponding states approach. LCP data was compiled or collected at several temperatures and QSPR predictions were made at each temperature point. Then, using statistical analysis to combine uncertainties, a modified corresponding states method was used to link the individual QSPRs while maintaining the known behavior as the critical point is approached.

This method improves on existing QSPRs for LCP by using unique descriptors and adding a temperature dependence that has not been previously implemented. This functionality gives process engineers a thermodynamically rigorous prediction method that can be used to extrapolate up to the critical temperature.

BYU

589934 Computational Tools for Physical Property Estimation and Solvent Selection

Abstract:

We will present the details of a computational, physical property-based solvent selection methodology that we have used in numerous process development projects. We will use the example of selecting a solvent for a liquid extraction process to explain the methodology's steps. Specifically, we will discuss:

- Identification of physical property constraints, e.g., the high solubility of the key component, immiscibility with the feed solvent, proper volatility for regeneration by distillation, etc.
- Identification of required or disallowed substructures, e.g., restricting chemically or thermally unstable substructures, requiring reactive substructures, restricting environmentally undesired substructures, etc.
- Estimation of required physical properties, e.g., liquid-liquid equilibrium for extraction, vapor pressures for distillation, surface tensions and densities for settling, etc.
- Screening and ranking of candidate solvents, e.g., selecting a database of candidate solvents, identifying general trends in recommendations, iterating on the entire selection process, etc.

We will demonstrate how this selection methodology has been implemented in a software tool and show how such software tools are essential for conducting a rapid and thorough selection process.

SwIFT (SwRI Interdependent Flexible Time-Driven) Process Development
Paper #181c

Monica Brinkman; Eloy Flores, III; Kyle March; Southwest Research Institute, San Antonio, TX

Abstract

For many new technologies and processes, the timeframe to go from bench-scale to production can take multiple years. Companies are often racing to build and develop a new technology with limited information about how their innovation will translate from their lab experiments to pilot units or full-scale production. Southwest Research Institute uses its SwIFT (SwRI's Interdependent Flexible Time-Driven) technique in pilot plant development for rapid design, fabrication, and testing of integrated pilot plant units to collapse the overall schedule for the project. SwIFT process development, similar to Agile Programming, uses several small design and redesign cycles throughout the project to increase feedback and uncover unknown risks and information as early as possible in the project cycle allowing them to be corrected in the first version. This greatly reduces the project schedule and allows for flexibility in design, fabrication, and operations as the design team discovers information regarding the new technology's implementation.

The SwIFT Process

SwRI's Interdependent, Flexible, Time-Driven (SwIFT) Process Development centers on quickly and efficiently brining new processes to commercialization. The SwIFT process focuses on the following factors:

- Variables and designs in the process are ***interdependent***. Design for unit operations and systems within the process are continuously being reviewed, reevaluated, and improved as new information is discovered.
- ***Flexibility*** is built into the design, so that changes can be made to the system when new information is identified or refined without rebuilding the entire unit.
- The project goals remain ***time-driven***. Schedule is critical for its effect on the budget, scope, delivery, and schedule constraints.

These concepts are combined to result in a honeycomb-like design-build structure. Design-build is a technique in which several small design, build, and test cycles operate in parallel instead of in series in a project in order to accelerate schedule. As each design-build-test cycle completes, more information is learned about the overall process and can be incorporated in other aspects and design cycles.

Step 1: Categorize Data

In the development of new technologies, there is a large amount of unknown information at the beginning of a project. At the start of the project, SwRI evaluates parameters needed for the design and categorizes the parameters as known information or unknown/assumed information. If there are any design-build-test cycles that can be completed with the known data, these are started at this time.

The way that SwRI does this is by generating simple engineering design documents such as input-output diagrams, block flow diagrams, and process flow diagrams. This creates a starting place for engineers to identify what data is needed and then begin to further categorize that data into known and unknown data.

Like other aspects of the SwIFT process, categorizing data is an iterative process that must often be revisited. As more data becomes known and the engineers can dive deeper into the design process, more unknown data may become apparent. For example, in a block flow diagram, it may be identified that a previously unrecognized separation process is needed. Initially, the type of separation process needed is unknown. Once the type of separation process required is identified, for example a crystallization unit, there are more unknowns around the design of the crystallization unit that have to be resolved such as nucleation rates, particle size and distribution, residence time, effect on stream composition, separation efficiency, and other process parameters.

Step 2: Minimize Unknown Data

Once the data is categorized, it is prudent to move as much of the unknown data as possible into the known data category. This is often done through as-needed laboratory testing.

SwRI evaluates any small-scale experiments or easily-tested properties to move the unknown/assumed information into the known category. Any tests with quick turnaround times, such as ASTM property tests or small-nonintegrated bench-scale tests are completed. This minimizes the unknowns and reduces the number of assumptions being made in the design, or worse, discovered in operation.

Step 3: Bound Unknown Data

Next, SwRI evaluates impact of the information that is unknown or has assumed values and bounds the information into a range with upper and lower limits instead of a single point. Engineers and scientists then design the pilot unit to handle the broad range of values so that the precise value to be used for plant design can be determined during pilot plant testing and operations. Where possible, flexibility is built into the test system to make it as robust as possible, such as using parallel components.

For example, based on screening tests, one may determine that a reactor will likely operate best at 475°C, 10 psig, and 1.5 seconds residence time. However, a full characterization of the product may not be able to be completed until the material is tested in a large-scale pilot unit, so optimization runs may need to be run once the pilot unit is complete. Additionally, the reactants may act differently in a larger unit or in an integrated system perhaps because, for example, of a temperature gradient in larger equipment. Perhaps new byproducts or transport issues not seen at the bench-scale screening test may be observed. By bounding design parameters and designing a flexible system that can operate between

400 – 600°C, at pressures 5-20 psig, and with a residence time of 1 – 3 seconds, the system now has flexibility to complete optimization runs at larger scale and to make changes in the process parameters without rebuilding equipment.

What if it is not feasible to run a screening test or lab-scale experiment? Sometimes, the complexity or cost of a screening test makes it impractical to run a bench scale test. If tests cannot be run, literature data might be used to supplement unknown design data and to guide where the bounding should be. Occasionally, depending on how far you are on the edge of technology, there is no information available at all and the lab-scale experiments give minimal information. In this case, multiple design paths may need to be considered, as discussed in the next section. Alternatively, correlation or theoretical estimates may suffice until operating measurements are made. Having equipment with enough leeway to accommodate this unknown can save refits and thus time on the schedule.

Step 4: Evaluate the Impact of Unknown Data

At this stage, SwRI determines the dependencies of the design information. Unknowns or assumptions that affect multiple parts of the pilot unit are given more weight/attention than those that have less impact on the overall process. Occasionally, for high risk or long lead time items that can only be tested in a fully integrated system, parallel design options can be fabricated so that they can be quickly and easily switched during testing.

Additionally, if a small part of the pilot unit can be built and tested, this is done concurrently with construction prior to full commissioning. Priority is generally given to the high-impact unknowns. Often, fabrication and testing of different pieces in the unit begin prior to completion of the full design with the focus on eliminating or validating as many of the unknowns/assumptions as possible. If an assumption proves incorrect, the design team will use the information gathered from the singular equipment testing to modify the design, a step more preferable than revising the finished pilot plant.

However, some assumptions can only be validated or disproved in a fully integrated system. In this case, the design and fabrication team must be prepared to respond quickly with a new design based on the information gathered during the early operations and testing period.

Step 5: Create a "Critical" Design Path

Prioritize which units in a pilot system need to be designed first and identify which are dependent on the designs/outputs of others. Traditionally, one might design/fabricate a system in the process flow order. For the SwIFT process design, after laying out the known and unknown data in a process flow diagram and evaluating the impacts of the unknown data, you would then generate a "critical" design path to determine the equipment design order. Once this initial "critical" design path has been identified, a P&ID is started concurrently with the unit operation/equipment design and a second more detailed "critical" design path is generated in parallel. This allows engineers to start design work on the large, long lead-time items in parallel with the detailed P&ID and other design work. Due to the complexity of the larger unit operations, final approvals for the single unit operations are ready in the time it takes to put together the design documents for the rest of the pilot unit.

Just like in scheduling, design has dependencies. By determining what design parameters are affecting the greater number of unit operations, one can determine which properties most urgently needed to be determined and calculate them first. In addition to which unit operations have been affected, lead-time for unit operations should also be taken into account when prioritizing design parameters. If parameter A affects more unit operations in total, but the unit operations only take a few days or a couple weeks to arrive, then it will have less impact than parameter B which only affects a single unit operation that may have a 12-week lead time.

Step 6: Continually Update the Design

The design can be continually updated during the design-build-test cycles; especially for items or design parameters with multiple dependencies. The design-build-test cycle is completed as many times as needed throughout the project for small and large items. Flexibility is key, particularly if the number of unknowns at the start of the project is large. Implementing a first-of-kind technology advances the "cutting edge" and will benefit from the SwIFT approach.

The SwIFT goal is to quickly produce a robust solution that can adapt to developing technical challenges as they are identified, shakedown and troubleshoot problems early as they arise, and bring the number of technical unknowns down to zero, derisking the client's process and making it ready for commercialization.

Advantages and Pitfalls of the SwIFT Process over Traditional Development

Although the SwIFT technique likely has more cycles of design-build-test, it allows the design team to discover and shakeout failures more quickly than a traditional prototype creation cycle. This rapid development approach can take years off a development cycle and get the company into production faster.

Advantages of the SwIFT Process

Time: While the traditional linear design process may take years to get to commercialization, the SwIFT Process Development generally starts to produce pilot-plant data that can be used for commercial design in 1-2 years with some pilot plants up and running in less than 6 months.

Data Density: Information is revealed quickly and efficiently. This allows the client to quickly make informed decisions about the process and the overall system, say including utilities or product handling. If a system has a fatal flaw, the client is in a better position, realizing this at month 6 rather than year 3. Stage gates for smaller tests often begin within a few months of the project start as data is generated even before the main fabrication of the pilot unit begins.

Flexibility: Since new information is quickly available, the team has the ability to adapt the design based on the new information as it becomes available. This sometimes completely changes the direction of the project, but allows for a greater chance of success.

Innovation: With a high data density and flexibility available, innovative solutions can be achieved for unique technical challenges. The iterative process may reveal aspects that were not evident in the process conception.

Pitfalls of the SwIFT Process

Cost: When rapid schedule and increased scope are higher priorities of the triple constraints of time, cost, and quality, cost will be unavoidably higher. Sometimes this is a matter of "pay me now or pay me later". The broader SwIFT approach avoids costly surprises that kill projects by resolving them early.

Aesthetics: Although safety is still the number one priority, rapid installation of quickly fabricated items is often "functional and neat, but not very pretty". For example, operators may grind down welds so that they will not cut any passersby, but not until they are polished enough to see your reflection. Another example may be when wire is being pulled or added to the control box because a new sensor was added to the system; it is safely in the cable tray but may not be bundled with cable ties since it may have been rerouted several times depending on how many times the design has changed.

Rapid Decision Making: The client and SwRI staff have to make quick decisions after running tests and analyses. In order to keep schedule requirements, there is precious little time to mull over decisions, analyze data, or change your mind. This may require more client engagement than originally considered.

SwRI Case Study – SwIFT Process Development for Gas to Liquids

SwRI worked with a client using the SwIFT Process Development for a pilot unit to produce olefins from natural gas. The main project goals were 1) rapid time-frame development of the pilot unit to make up for time lost on the schedule prior to starting the project at SwRI, and 2) a proven inline analytical system. To help expedite the process, the pilot skid was built at SwRI at a "brown site", a site which already had utilities, which previously had been used for a different technology platform.

For the first step, identifying unknowns, SwRI engineers and operators first keyed-in on the site work and utilities. Many of the skids, such as a molecular sieve, natural gas, and compressor skids, had not been used for several years and needed to be recommissioned or upgraded. Additionally, for this pilot unit, new systems requiring oxygen-clean equipment would need to be added. Another unknown, which was a top priority for this client (and most of our clients) was the analytical system setup. For pilot plants, the process data and the analytical data are the outputs used to evaluate the process and design for the subsequent commercial step. The client needed inline analytical data that would tie to the process data. Although there were many other unknowns, a fourth large unknown was heat transfer and heat loss in the system. Both the client and SwRI modeled the heat transfer, but these needed to be validated and tested in hardware to ensure that the temperature range for testing could be reached.

In the second step, minimizing unknowns through testing, SwRI began to complete initial testing on components where possible. For starters, an initial heater function test was completed on the methane and nitrogen heaters in order to validate the heat-up rate, heat-up duration, and ability to reach the set point temperature at the desired flow rates. Although the initial model suggested the heat of reaction would provide sufficient heat to the system, a larger internal heater had to be installed during the testing process to make up for heat losses throughout the unit. Additionally, onsite testing was done to find a special high-temperature seal appropriate for the application. In the reactor itself, insulation types were tested in addition to the validation of the heat loss models. With each run, improvements were made to the design until a three-layer insulation design was finalized, which allowed the reactor to maintain the desired run temperature. Another challenge, which included several unknowns, included the inline analytical system. There was a great distance from the reactor to the analytical instrument, which could greatly impact the temperature profile in the analytical lines. Additionally, water entrainment was a concern. These problems were addressed as encountered yielding a faster result.

For the third and fourth steps, bound unknowns and evaluate impact, engineers sized equipment, such as flowmeters, in the system to give the most flexibility. However, even though the unit operations allowed for a high degree of flexibility, the final operating regime ended up drastically different, and one of the heat exchangers required a redesign. Since engineers were constantly monitoring the impact of changes in the system, the project team was prepared for the redesign and was able to quickly turnaround the design, fabrication, and installation to get the system back up and running during operations.

For the fifth step, create a critical design path, the project team ordered long lead time items first, and made good use of long lead times by working many projects in parallel. When considering which items to finalize in the design and to order first, priority was given to unit operations which were found to have the longest lead times. Additionally, for specialty equipment such as the oxygen-clean equipment, the training and site preparation for installation were completed faster in-house (compared to awaiting a contractor) while awaiting equipment delivery.

Also, as part of the critical design path, there were several operations within the project that were scheduled in parallel. The natural gas skid recommissioning occurred at the same time as the Hazards and Operability (HazOp) review since the natural gas skid had been previously operated, had its own existing HazOp, and there were no modifications made to it. The molecular sieve media replacement, compressor upgrade, and reactor design were all completed simultaneously. Controls programming and system buildup occurred at the same time. Analytical commissioning occurred while the wiring and functional testing of instrumentation was being completed. All of these allowed the project to safely compress schedule.

As a result of using the SwIFT Process Design method, SwRI engineers and scientists were able to condense this 15-month build-up of the system to 6 months. During operations, SwRI engineers and scientists successfully tested multiple catalysts, multiple reactor inserts, reactor supports, heat retention, and validated the heat balance. There was room for modifications throughout operations as new information was learned about the process and SwRI engineers were able to quickly adapt the pilot system to generate the data needed for the client.

Conclusion

Overall, SwIFT Process Development reduces schedule, reveals unknowns in new systems earlier in the development process, discovers potential technical missteps or misunderstandings at an early stage, and saves money in the long run, although more upfront cost is required at the start for testing and to build flexibility into the system.

Using the SwIFT Process Development for clients, SwRI has greatly reduced the time for clients to move their process from bench-scale to commercialization, with project timelines ranging from 6 months to 3 years. SwRI's 1500 acre site in San Antonio, Texas is uniquely suited to pilot plant design, buildup, and operations having adequate buffer space and operating safety systems. Southwest Research Institute possesses pilot plant emission permits, multiple fabrication and machine shops for rapid construction and modification, 24/7 operations and analytical services available, and multidisciplinary expertise for any unexpected technical challenges in engineering, chemistry, analysis, automation, fabrication, or operations.

Standardizing Risk Management in Laboratory and Pilot Plant Facilities: A Model

Karen F. Carter
Process Safety Team Lead
Shell Technology Center Houston
k.carter@shell.com

(Presentation notes and additional information slides have been added to this file)

The Ask

- What hazards on our site have higher levels of risk (aka Yellow risks)?
- Where are these Yellow risks located?
- Document the barriers used to manage these Yellow risks.
- Obtain Management agreement that the risks in their operations are managed to ALARP.

The Challenge

(Do it in one year!)

Shell Technology Center Houston:

- >150 labs and pilot units
- >10,000 different chemicals

Research and development activities include:

- Analytical processes/services
- Catalyst development/testing
- Hydrocarbon sample and product testing
- Process and technology development
- Work with new/alternative energies, fuels, lubricants, catalysts
- Pilot plants

We run the gamut from extremely small scale to medium scale operations. We're not at production level that you would have in a manufacturing location, but there is immense variety in the types of experimental operations at our site. How can one program cover such a wide range of businesses?

Documenting ALARP for Yellow Risks

Hazard Register has list of yellow risks and how to document ALARP. ID your hazards and locate them on the Hazard Register.

H-phrases and H-codes for each hazard will point to the Model BowTies used to identify the <u>minimum</u> requirements for ALARP.

Format is provided to document the Hazards, Minimum Barriers, plus additional barriers used for ALARP Determination.

Hazard Register = Hazard and Effects Register
As Low As Reasonably Practicable

What do we already have? HAZID methods

H-phrase	Hazard from SDS
H220	Extremely flammable gas
H221	Flammable gas
H222	Extremely flammable aerosol
H223	Flammable aerosol
H224	Extremely flammable liquid and vapor
H225	Highly flammable liquid and vapor
H226	Flammable liquid and vapor
H227	Combustible liquid
H228	Flammable solid
H229	Pressurized container: may burst if heated

H-code	Hazard from Manufacturing
H-25	**Ergonomic Hazards**
H-25.01	Repetitive Motion
H-25.02	Awkward Posture
H-25.03	Physical Effort (push, pull, lift, carry)
H-25.04	Sustained Posture (Maintaining Static Position)
H-25.05	Contact/Compression Stress
H-25.99	Other
H-26	**Physical Agents**
H-26.01	Noise
H-26.02	Heat Stress
H-26.03	Cold Stress
H-26.04	Vibration
H-26.05	Lighting

We have a large list of hazard types used in Manufacturing locations for hazard identification.

What do we already have? A Risk Assessment Matrix

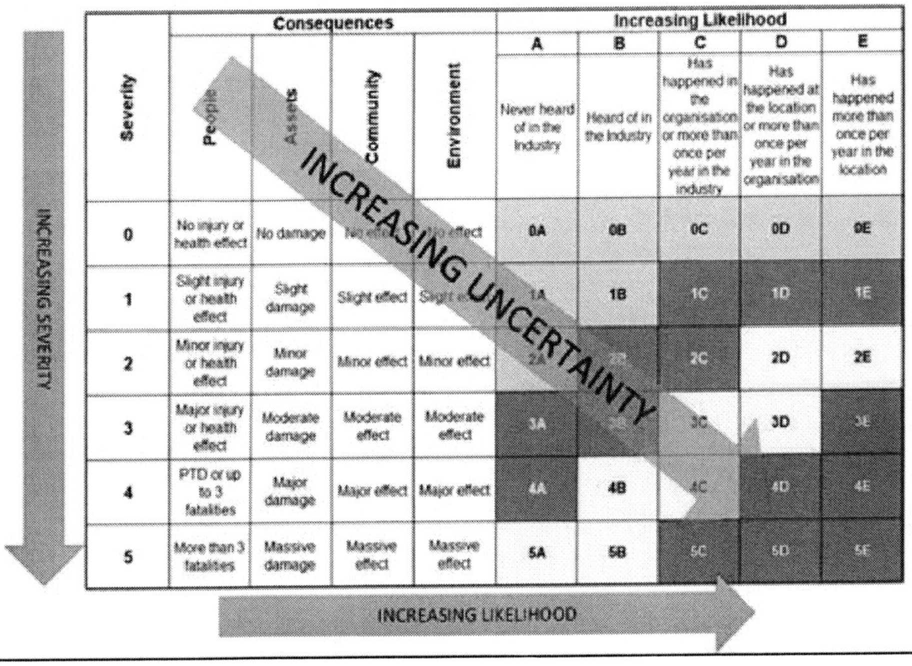

Severity	Consequences				Increasing Likelihood				
	People	Assets	Community	Environment	A	B	C	D	E
					Never heard of in the Industry	Heard of in the Industry	Has happened in the organisation or more than once per year in the industry	Has happened at the location or more than once per year in the organisation	Has happened more than once per year in the location
0	No injury or health effect	No damage	No effect	No effect	0A	0B	0C	0D	0E
1	Slight injury or health effect	Slight damage	Slight effect	Slight effect	1A	1B	1C	1D	1E
2	Minor injury or health effect	Minor damage	Minor effect	Minor effect	2A	2B	2C	2D	2E
3	Major injury or health effect	Moderate damage	Moderate effect	Moderate effect	3A	3B	3C	3D	3E
4	PTD or up to 3 fatalities	Major damage	Major effect	Major effect	4A	4B	4C	4D	4E
5	More than 3 fatalities	Massive damage	Massive effect	Massive effect	5A	5B	5C	5D	5E

We know how to do HAZIDs, and we can rank the risk of hazards.

What do we already have? A Hazard Register

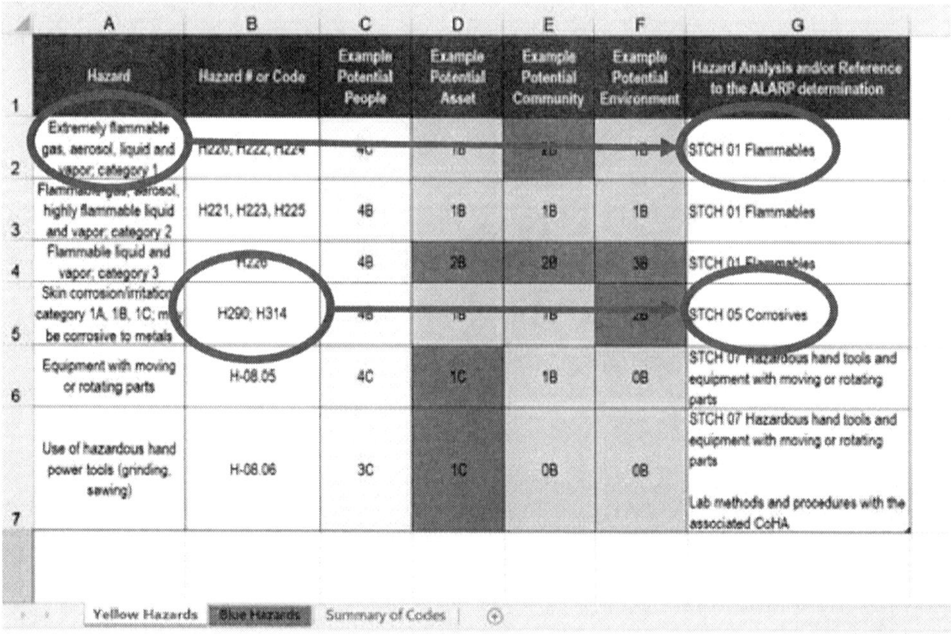

	A	B	C	D	E	F	G
1	Hazard	Hazard # or Code	Example Potential People	Example Potential Asset	Example Potential Community	Example Potential Environment	Hazard Analysis and/or Reference to the ALARP determination
2	Extremely flammable gas, aerosol, liquid and vapor; category 1	H220, H222, H224	4C	1B	2B	1B	STCH 01 Flammables
3	Flammable gas, aerosol, highly flammable liquid and vapor; category 2	H221, H223, H225	4B	1B	1B	1B	STCH 01 Flammables
4	Flammable liquid and vapor; category 3	H226	4B	2B	2B	3B	STCH 01 Flammables
5	Skin corrosion/irritation category 1A, 1B, 1C; may be corrosive to metals	H290, H314	4B	1B	1B	2B	STCH 05 Corrosives
6	Equipment with moving or rotating parts	H-08.05	4C	1C	1B	0B	STCH 07 Hazardous hand tools and equipment with moving or rotating parts
7	Use of hazardous hand power tools (grinding, sawing)	H-08.06	3C	1C	0B	0B	STCH 07 Hazardous hand tools and equipment with moving or rotating parts Lab methods and procedures with the associated CoHA

Yellow Hazards | Blue Hazards | Summary of Codes

Either use the Hazard Description to determine the BowTie to be used (**Brainstorm your hazards**)
Or use **H-Phrases from SDS and H-Codes from our hazard lists** to look up applicable hazards, that will determine the BowTie to be used (Specific hazard lookup)
We still needed the Model BowTies that had the minimum required barriers to reduce the risk from each hazard to ALARP

We needed: the Model BowTies

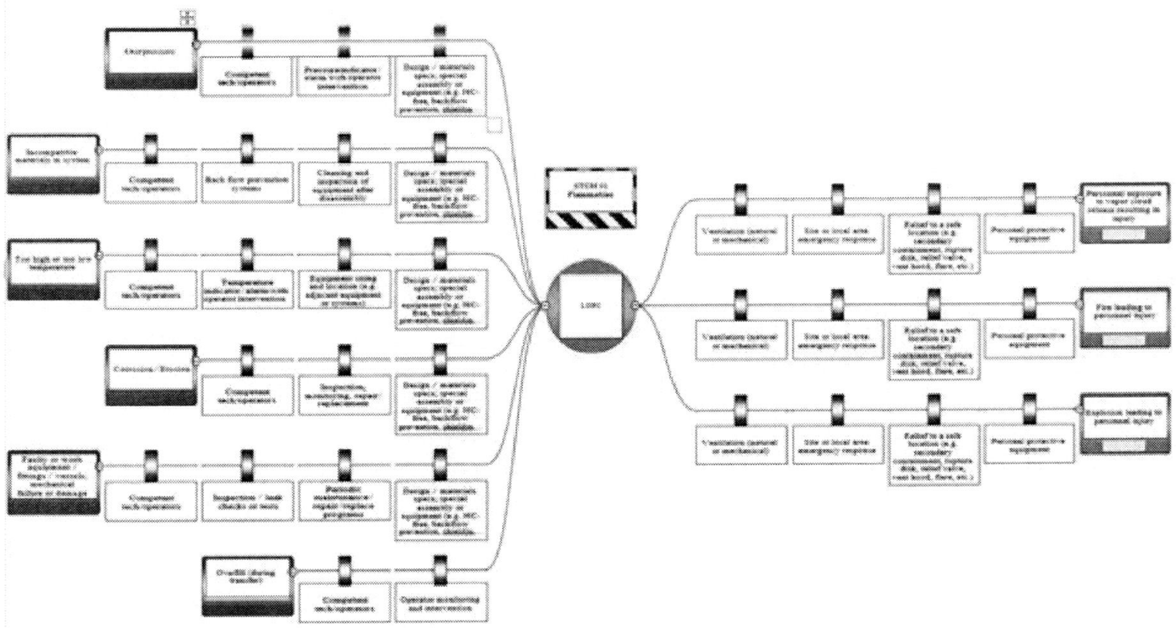

Model BowTies have the MINIMUM required barriers to reduce the risk from a hazard to ALARP

New Model BowTies

1. Flammables
2. Pyrophoric/water reactive
3. Oxidizer
4. Gas/Process Pressure
5. Corrosive
6. Acute Toxic
7. Moving/Rotating Parts
8. Work at Height > 1.8m
9. Work at Height < 1.8m
10. Overhead equipment
11. Work Below Grade
12. Objects under Induced Stress
13. Knives/Sharps
14. Heat Stress
15. Temp > 150F
16. Open Flame
17. Electricity

Initial ALARP Determination Process

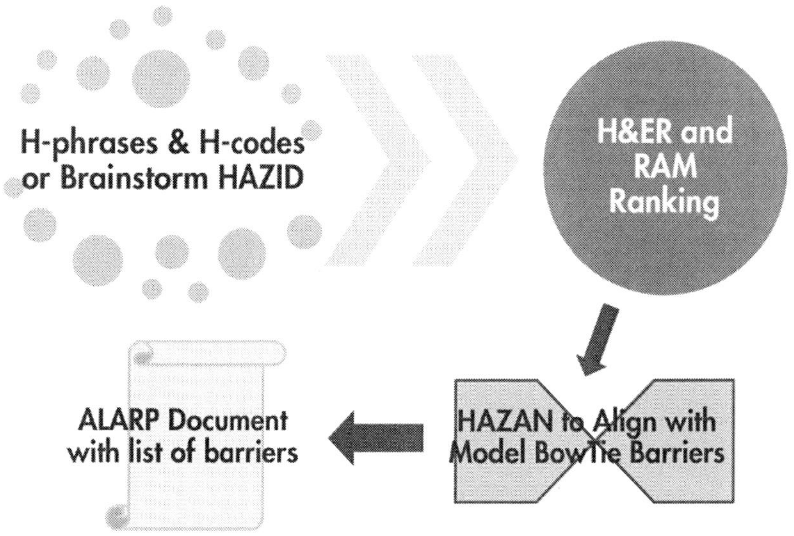

- HAZID, analyze/rank risks, determine Model BowTie barriers, additional barriers, and gaps (if applicable); prepare ALARP documentation for Manager's acceptance.
- At ALARP means all barriers are in place and operational; otherwise a gap closure plan is documented.
- This becomes the **baseline** for all future changes

ALARP Determination & Management Approval

ALARP Determination Form for RAM Yellow Hazards
(not to be used for RAM 5A/B Yellow or Red Hazards)

[redacted]	**Lab Manager:**	**XXX**	**Assessment Completed:**	2/18/2020
	Approval:	**YYY**	**Approval Date:**	2/18/2020

Hazard	Present as a Yellow Risk?	Specific Description of the Hazard
Flammables	yes	process gas (ethylene, EO, ethane) ethylene ethane hydrogen Isopar-H cal gas (ethylene & methane)
Gases/processes under pressure	yes	reactors (process gas, water, Isopar-H) pressure vessels guard beds cylinders (H2, N2, EO/N2, CH4, C2H6, He, cal gasses) flex hoses Dracite tubes
Acute toxicity	yes	ethylene oxide vinyl chloride ethyl chloride
Power tools and equipment with moving or rotating parts	yes	fans compressors
Personnel working at height > 6 feet	yes	roof work ladders elevated platforms
Personnel working at height < 6 feet or at grade	yes	3-step ladders
Electricity	yes	120 V+ circuits

Hazard	Threat	Barrier	Specific Description of Barrier	Who Implements and Maintains this Barrier?
Flammables Gases/processes under pressure Acute toxicity	Overpressure	Competent tech/operators	Hands-on training for new staff. Required annual reading of SOP.	Lead Operator responsible for staff training.
Flammables Gases/processes under pressure Acute toxicity	Overpressure	Pressure indicator / alarm with operator intervention	All reactors, six packs, feed lines, reactor outlets have pressure gauges or transmitters. All transmitters tied into the DCS and have lowP and highP alarms. Alarms are audible & visible, and go to Security Control after hours. High-pressure shutdown on jacket pressure for EO-1 and EO-2.	WICS calibrate pressure transmitters & shut-down testing every 30 months.
Flammables Gases/processes under pressure Acute toxicity	Overpressure	Design / materials specs, special assembly or equipment (e.g., HC-free, backflow prevention, shielding)	All equipment designed to handle standard operating pressures	MOC process determines if changes meet SHE requirements.
Flammables Gases/processes under pressure Acute toxicity	Incompatible materials in system	Competent tech/operators	Hands-on training for new staff. Required annual reading of SOP.	Lead Operator responsible for staff training.
			Check valves on all feed lines (before and after guard beds) to	

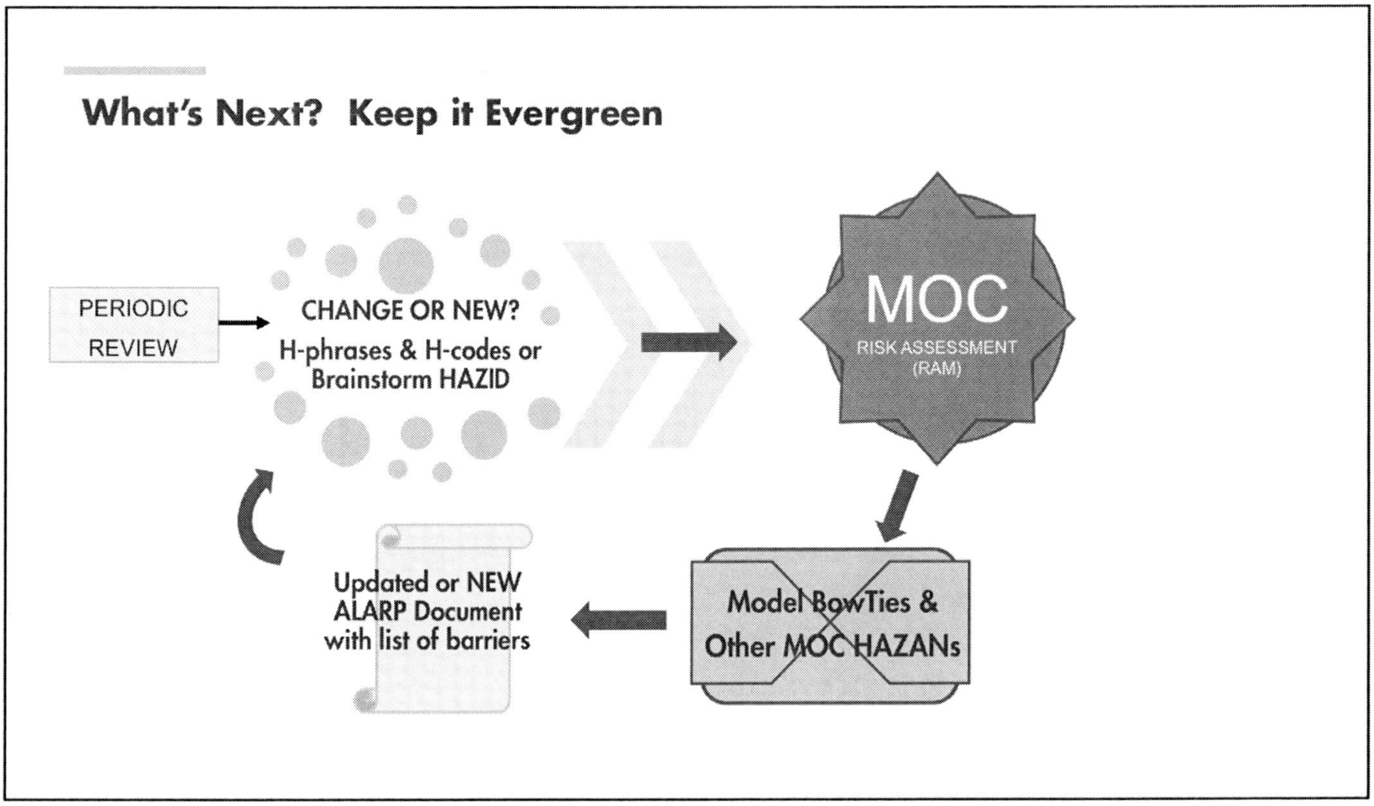

- NO CHANGE TO A PROCESS? Periodically revalidate ALARP documentation (every 5 years at our site).

- MOCs will change existing ALARP Determinations. Barriers and hazards may be added or removed; entire processes may be removed, changed, or moved. If MOC is used to assess changes and additions to a lab/plant, part of the MOC process is to require updating the ALARP Determination for that lab/plant.

- Model Bowties are no longer the ONLY way to determine if a RAM Yellow Risk is at ALARP, it is just one methodology available. An MOC program may have a variety of hazard assessment methodologies that can be used to determine ALARP, but the Model BowTie method is still flexible, easy to use, and fit for purpose in an experimental operations environment.

Questions and Answers

Lessons Learned

Interesting fact – chemical BowTies had *almost all* identical barriers

RAGAGEP became ***Recognized and Generally Accepted Good Lab Practices - RAGAGLP***

- ❑ Chemical Hygiene Plan
- ❑ PPE in the Certificates of Hazard Assessment
- ❑ Codes and standards that regulate physical hazards. e.g.
  - Working at heights
  - Rotating equipment
  - Electricity

A BowTie can't replace a brain!

Common barriers were mostly the recovery measures
Some chemicals with very different control barriers were in asphyxiants, highly toxics, highly reactive agents

You must review each step of this process to confirm it's correct and complete. It's not a no-brainer.

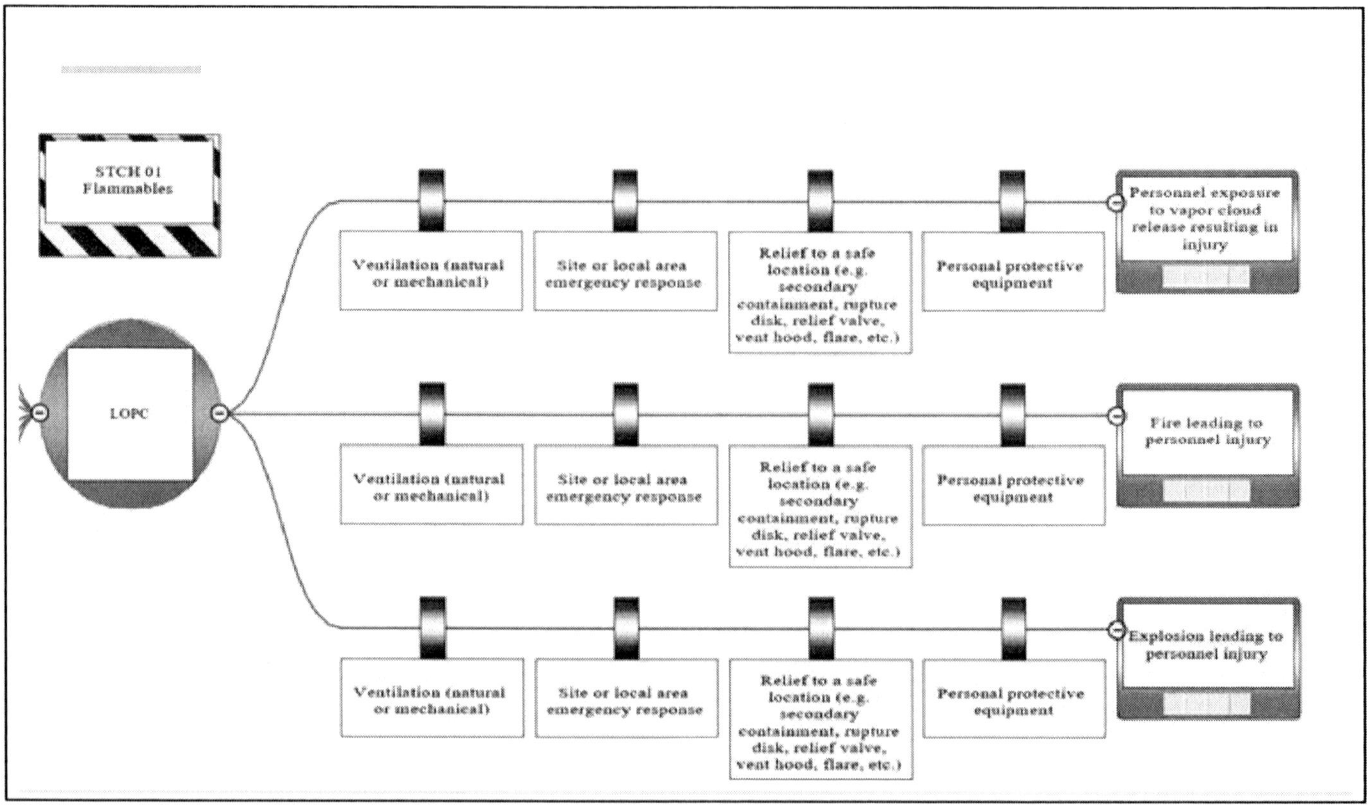
STCH 01
Flammables
LOPC
Ventilation (natural or mechanical)
Site or local area emergency response
Relief to a safe location (e.g. secondary containment, rupture disk, relief valve, vent hood, flare, etc.)
Personal protective equipment
Personnel exposure to vapor cloud release resulting in injury
Ventilation (natural or mechanical)
Site or local area emergency response
Relief to a safe location (e.g. secondary containment, rupture disk, relief valve, vent hood, flare, etc.)
Personal protective equipment
Fire leading to personnel injury
Ventilation (natural or mechanical)
Site or local area emergency response
Relief to a safe location (e.g. secondary containment, rupture disk, relief valve, vent hood, flare, etc.)
Personal protective equipment
Explosion leading to personnel injury

Abstract

By its very nature, a research and development site is in constant change. We try things that have not been done before. We create and test new processes, new chemical formulae, new products and new production techniques. Is it possible to find common ground for managing risk in such a diverse environment? The answer is yes.... But. With careful assessment of the risks and acknowledgement when the risks are truly unknown, a standard for the minimum level and types of acceptable controls and barriers can be established. But, that is a minimum. There is still some work to do to reduce your risk to as low as reasonably practicable. See how the Shell Technology Center Houston has used a risk assessment matrix and model BowTies to develop a process to standardize the minimum requirements needed to manage health, safety and environmental risks.

591119 Managing a Safety Culture throughout Mergers and Acquisitions

Abstract:

Mergers and acquisitions are common in the chemical industry; approximately $50 billion in M&A deals are executed each year. While the focus from the corner office is on the financial aspects of the consolidation, the reality is that when the ink is dry two different groups will combine their processes, cultures, and approaches to safety into a single entity.

This presentation discusses how one company manages its safety and compliance objectives throughout the M&A process. First-hand perspectives will be presented from the viewpoints of both the parent company and the acquisition target. Topics include: how to merge two separate approaches to safety, potential pitfalls to stay vigilant against, best practices before, during, and after the acquisition.

590420 Cubic-Plus-Chain (CPC): A SAFT-Based Chain Modification to the Cubic Equation of State for Large Nonpolar Molecules

Abstract:

An equation of state framework for nonpolar chain molecules is proposed that hybridizes the classical cubic equation of state with the chain equation of state from SAFT. The cubic equation of state serves as the physical description of the monomer beads, providing repulsive and attractive character to the segments, and the chain term bonds these monomer segments to form chains of homogeneous beads. The CPC model molecule is a better representation of chain-like molecules, such as *n*-alkanes and polymers, and phase behavior modeling with the proposed CPC equation of state shows considerable improvement over the reference cubic equation of state. Additionally, CPC compares well to PC-SAFT in its description of both hydrocarbon VLE and polymer/solvent LLE despite being substantially less computationally expensive. The polymer/solvent example illustrates that a relatively simple modification can be made to the classical cubic equation of state that significantly improves density predictions while retaining many of the properties of the classical cubic equation of state that make it a popular model.

590322 Developing Reactor Scale Kinetics Model Integrated with Hydrodynamics Towards Scale-up of a Packed Bed Reactor

Abstract:

Assessing the reactor performance for trickle bed and two-phase upflow packed bed reactors needs optimized experimentation supplemented by validation of a reliable mathematical model at different scales. Various factors like catalyst wetting, liquid maldistribution, and transports are necessary to predict the performance of these reactors. At Multiphase Flows and Reactors Engineering and Applications laboratory (mFReal), a lab-scale packed bed reactor experimental unit equipped with a blast shield has been designed to conduct flow reaction investigations at high-pressure and temperature. The unit was tested for selected operating conditions for the acetylene hydrogenation in the liquid phase to understand the conversion and optimize the process towards scaleup aided with the model. Reactor scale models have been developed towards this study based on different approaches incorporating the kinetics and hydrodynamics with the effectiveness factor evaluated from a pellet scale model.

Catalytic hydrogenation of acetylene in the gas phase to form ethylene has widely been used in industrial practice. This reaction is highly exothermic which may lead to thermal runaway. Moreover, many studies have reported the formation of green oil (C_6^+) due to the oligomerization of the reaction leading to catalyst deactivation. The hydrogenation of acetylene in the liquid phase using a selective solvent is a study of interest to overcome the issues of green oil formation and controlling the heat of the reaction. The solvent used in this study is N-Methyl Pyrrolidone (NMP), a polar solvent generally used to strip acetylene selectively from the cracker effluent. The reaction kinetics of the acetylene hydrogenation over a commercial 0.5 wt% Pd/Al_2O_3 catalyst was investigated in a stirred tank reactor. The reaction was studied for various operating conditions to mention temperature, catalyst loading at a pressure of 250 psig. The kinetic parameters were estimated by Langmuir-Hinshelwood-Hougen-Watson based reaction mechanism by fitting the model equation to the experimental data. The kinetics of the reaction will be integrated into the model to predict the reactor performance of a packed bed reactor at scale. Cold flow experiments in a trickle-bed reactor at scaled-down operating conditions using a liquid tracer were conducted to obtain the residence time distribution (RTD). The operating conditions were chosen from literature and industrial conditions. From the RTD information, the liquid dispersion and liquid holdup were evaluated. Integrating both the kinetics and hydrodynamics, the reactor scale model was enhanced and validated against experimental data obtained from the flow hydrogenation experiments conducted at optimized operating conditions. This information will help us to evaluate and suggest ideal scale-up parameters for this reaction. This study can be expanded to various sensitive catalytic reactions.

590295 Advantages and Drawbacks of Coarse-Grained CFD-DEM and Eulerian Multiphase Simulation for Plant-Scale Fluidized Beds

Abstract:

Since its introduction [1], the Discrete Element Method (DEM) has proven to be a valuable method for the analyzing and understanding particulate flows. Supported by the continuously increasing computational power, CFD-DEM simulations have found their way into the chemical and process industry for various applications like solid suspension in mixing vessels, fluidized and spouted beds, granular transport and coating applications in rotary drums [2,3].

The major shortcoming of DEM, however, is its computational cost that increases with the amount of particles involved, their material properties (stiffness) and size. This hinders the application of CFD-DEM simulation to large-scale systems of industrial size. To overcome this shortcoming a coarse grain (CG) model has been described [4]. Using straightforward scaling rules, a group of particles gets replaced by a representative coarse parcel. This effectively reduces the number of particles that need to be processed and subsequently shortens the computational time. On the other hand they introduce a modeling error into the simulation. In a previous study we were investigating different coarse graining models and quantify the effect on different parameters and their fluctuation like pressure drop and expansion height in a fluidized and spouted bed and we were comparing the simulation with experimental results. In general it can be stated that bed expansion and pressure drop are well predicted for all investigated CG methods, but depending on the CG factor the dynamic of the system cannot be captured. [NEU]

In this work we are comparing these CFD-DEM results with simulation results from a particle unresolved simulation based on the Kinetic Theory of Granular Flow (KTGF). In this modeling framework the different phases are treated as interpenetrating continua (Eulerian Multiphase), but additionally the intraphase interaction of the dispersed solid phase is considered. A validation of different KTGF-models and the influencing model parameters is discussed in lab-scale and results for industrial scale are shown. Additionally the KTGF results are compared against the DEM results w.r.t. computational cost and accuracy.

References:

1. A. Cundall, O.D.L. Strack, Geotechnique, 29 (1979), 47
2. Eppinger, O. Baran, R. Aglave, S. Lo, Simulating Solid Suspension in Stirred Vessels with a Fully Coupled CFD-DEM Algorithm (161d), AIChE Annual Meeting 2017, Minneapolis
3. Baran, R. Aglave, M. Tandon, A. Karnik, S. Lo, Numerical Simulation of Dense Gas-Solid Fluidized Beds: Comparison Between Eulerian Multiphase and Discrete Element Methods, AIChE Annual Meeting 2015, Salt Lake City, UT.
4. Sakai, Y. Yamada, Y. Shigeto, K. Shibata, V.M. Kawasaki, S. Koshizuka, Numerical Methods in Fluids 64 (2010), 1319
5. Eppinger, T., Tourani, C., Aglave, R., Quantifying Effects of Coarse-Grain Model Parameter in CFD-DEM Simulations of Fluidized and Spouted Beds and Comparison Against Eulerian Multiphase Simulation Results, AIChE Spring Meeting 2019, New Orleans

589932 Work and Heat Interconversion By Unicursal Cycle Processes

Abstract:

In general, an energy and material production process is a reciprocal or unicursal cycle process that starts and ends at environmental condition, in which work and heat are exchanged between process fluid and other systems such as heat generation system, waste heat removal system, etc. The first law gives the input-output energy balance. Work and heat are interconverted through internal energy of system in the unicursal cycle process which is composed of four thermodynamic elemental process modules: (1) isobaric heating process module (heat receiver, HR), (2) adiabatic compression process module (work receiver, WR), (3) isobaric cooling process module (heat transmitter, HT) and (4) adiabatic expansion process module (work transmitter, WT). Four different combination of thermodynamic elemental process modules as cycle processes can be considered; heat engine, heat pump, hot and cold heat cogenerator and thermal motor. It is noteworthy that no exergy destruction takes place in any thermodynamic elementary process modules. This means that both of exergy and anergy are conserved in the cycle processes.

In this study the work and heat interconversion mechanism of unicursal cycle processes is clarified in terms of exergy analysis. The conceptual design of thermal motor will be described.

590051 A New Kinetic Model for the Simulation of the Claus Process

Abstract:

The modified Claus process disposes of H_2S in acid gas streams from refinery amine units and sour water strippers, and converts it to elemental sulfur. The feed mixes with air or oxygen and is sent to a burner/waste heat boiler where combustion and other reactions occur transforming H_2S to SO_2 and sulfur. The outlet is sent to a series of catalytic fixed beds where most of the remaining H_2S reacts with SO_2 to produce elemental sulfur and water (Claus reaction). First bed operating conditions and catalyst are selected to favor the hydrolysis of CS_2 and COS formed as byproducts in the furnace. Sulfur production in the catalyst bed is favored at low temperatures with each bed operating at a slightly lower temperature. Sulfur is extracted by inter-stage condensation and separation. Tail gas out of the final bed can be sent to a tail gas unit to achieve overall sulfur recovery > 99%.

The focus of this work is the formulation and application of a new kinetic model for the Claus catalytic fixed bed reactors. The proposed reaction mechanism accounts for surface adsorption of intermediate species in line with experimental data on observed surface species concentrations. Experimental data on Claus reaction, hydrolysis of CS_2 and COS on alumina and titania catalysts at different temperatures and feed compositions were used to estimate kinetic and adsorption parameters for each type of catalyst using available gradient-based algorithms in the Optimization Tool^{TM} of the ProMax® simulation software.

The kinetic rate expressions and parameters were included in a model for a commercial sulfur recovery unit created using ProMax. Different cases were evaluated to study the effect of variations of feed composition, reactor temperatures and titania content of the first catalytic bed on sulfur recovery, with the goal of maximizing sulfur recovery.

AUTHOR INDEX

AIChE
120 Wall Street, FL 23
New York, NY 10005-4020

ISBN 978-1-7138-2345-2

9 781713 823452

CURRAN ASSOCIATES INC.
proceedings.com